Ergonomics in the Automotive Design Process

Automotive design continues to evolve at a rapid pace. As electric cars become ever more commonplace on the roads to the advent of the driverless vehicle, understanding the ergonomics behind automotive engineering becomes ever more paramount. Vehicle attributes must be considered early during the new vehicle development program by coordinated work of multi-disciplinary teams to begin creating vehicle specifications and development of vehicle attribute requirements.

In *Ergonomics in the Automotive Design Process: Advanced Topics, Measurements, Modeling and Research*, experienced automotive engineer Vivek D. Bhise investigates the advanced procedures and considerations to develop an ergonomic vehicle This book covers the entire range of ergonomics issues involved in designing a car or truck and offers evaluation techniques to avoid costly mistakes and assure high customer satisfaction. This book delves into driver performance, electric vehicles (EVs), interfaces, new technology and costs and benefits plus a lot more. Evaluation and measurement are covered in essential detail and the title has been brought right up to date with chapters on engineering design during automotive product development, vehicle evaluation, verification and validation and product liability litigations and ergonomic considerations.

This book is designed to allow the reader to develop a more comprehensive knowledge of issues facing the developers of automotive products and delivers methods to manage communication, coordination and integration processes. Delivering a toolkit that will allow you to implement systems engineering to minimize the risks of delays and cost overruns, it delivers a framework that will allow you to create the right product for your customers. The reader will therefore develop a knowledge of future in-vehicle devices that are easy to program and use, safe, cheap to manufacture and assemble and eco-friendly. This title is an ideal read for students and practitioners of ergonomics, human factors, automotive design, civil engineering, product design, work design and mechanical engineering.

Vivek D. Bhise is currently a LEO Lecturer/Visiting Professor and a Professor in post-retirement of Industrial and Manufacturing Systems Engineering at the University of Michigan-Dearborn. He received his B.Tech. in Mechanical Engineering (1965) from the Indian Institute of Technology, Bombay, India, M.S. in Industrial Engineering (1966) from the University of California, Berkeley, California and Ph.D. in Industrial and Systems Engineering (1971) from the Ohio State University, Columbus, Ohio. During 1973 to 2001, he has held several management and research positions at the Ford Motor Company in Dearborn, Michigan.

Ergonomics in the Automotive Design Process

Advanced Topics, Measurements, Modeling and Research

Volume 2

Second Edition

Vivek D. Bhise

CRC Press
Taylor & Francis Group
Boca Raton London New York

CRC Press is an imprint of the
Taylor & Francis Group, an **informa** business

Designed Cover Image: Vivek D. Bhise

First published 2024
by CRC Press
2385 NW Executive Center Drive, Suite 320, Boca Raton FL 33431

and by CRC Press
4 Park Square, Milton Park, Abingdon, Oxon, OX14 4RN

CRC Press is an imprint of Taylor & Francis Group, LLC

© 2024 Vivek D. Bhise

ISBN: 9781032739137 (hbk)
ISBN: 9781032779652 (pbk)
ISBN: 9781003485605 (ebk)

DOI: 10.1201/9781003485605

Typeset in Times
by Newgen Publishing UK

Access the Instructor and Student Resources/Support Material: www.routledge.com/9781032739137

Contents

Chapter 7

Preface

This new second volume of this book was created to provide additional material needed by ergonomics engineers to expand their effectiveness in implementation of ergonomics in future automotive products. It covers advanced topics such as prediction of visibility of targets and glare to evaluate vehicle lighting systems, driver performance and workload measurements, new technology implementations in creating advanced driver assistance systems, electric vehicles and autonomous vehicles. It also includes material to help implement ergonomics during detailed engineering design phase, management of interfaces between different vehicle systems, designing for special driver and user populations, methods for vehicle evaluation in verification and validation, models for conducting cost-benefit analyses, and involvement of ergonomics engineers during litigations associated with product liability cases.

It is hoped that this new Volume 2 will enable practitioners and students to understand the expanded scope of ergonomic considerations and inputs in the development of future automotive products and makes the two volume set of books more comprehensive and useful.

I would like to thank Suriya Rajasekar from Newgen KnowledgeWorks Pvt Ltd and Stuart Murray, Ann Chapman and James Hobb from Taylor and Francis for providing many valuable suggestions and guidance in creating this set of two volume books.

Vivek D. Bhise
March 13, 2024

Preface to the Second Edition

In this new edition, I have added nine new chapters to expand coverage of ergonomic considerations in the design of automotive products. I have also reorganized some of the topics, adding more material in several existing chapters to cover many ergonomic considerations for new features incorporated in vehicles since the first edition of this book. And some material that is no longer relevant was eliminated. While this edition maintains, the two parts contained in the original book, this new edition is now produced as a two volume set (Volume 1: Ergonomics Concepts, Issues and Methods in Vehicle Design [Chapters 1 to 13] and Volume 2: Advanced Topics, Measurements, Modeling and Research [Chapter 14 to 26]), new chapters are added to provide a more up-to-date coverage of new automotive features, issues and techniques needed to develop new complex automotive products with added emphasis on systems engineering process and other attributes affecting ergonomics such as packaging, comfort, safety, performance, emissions, manufacturing and assembly efficiency and sustainability.

The new chapters cover the following topics: 1) Ergonomics in the Systems Engineering Process [Chapter 2], 2) Decision Making and Risks in Automotive Product Programs [Chapter 3], 3) Ergonomic Considerations in Electric Vehicle Development [Chapter 17], 4) Ergonomic Issues in Autonomous Vehicles [Chapter 18], 5) Understanding Interfaces Between Vehicle Systems [Chapter 20], 6) Detailed Engineering Design During Automotive Product Development [Chapter 21], 7) Vehicle Evaluation, Verification and Validation [Chapter 22], 8) Costs and Benefit Considerations and Models [Chapter 23], and 9) Product Liability Litigations and Ergonomic Considerations [Chapter 26]. I have also included examples to elaborate several topics in key areas and applications included in many chapters.

The new revised edition, thus, is intended to provide practitioners and students in ergonomics a more comprehensive knowledge on issues facing the developers of automotive products and management personnel in communicating, coordinating and integrating vehicle development processes. It provides more tools in implementing the systems engineering to minimize the risks of delays and cost overruns, and more importantly, creates the right product for its customers.

Vivek D. Bhise
July 20, 2023

Preface to the First Edition

The purpose of this book is to provide a thorough understanding of ergonomic issues and to provide background information, principles, design guidelines, tools and methods used in designing and evaluating automotive products. This book has been written to satisfy the needs for both students and professionals who are genuinely interested in improving the usability of automotive products. Undergraduate and graduate students in engineering and industrial design will gain an understanding of the ergonomics engineer's work and the complex coordination and teamwork of many professionals in the automotive product development process. Students will learn the importance of timely information and recommendations provided by the ergonomics engineers and the methods and tools that are available to improve user acceptance. The professionals in the industry will realize that the days of considering ergonomics as a "commonsense" science and simply "winging-in" quick fixes to achieve user-friendliness are over. The auto industry is facing tough competition and severe economic constraints. Their products need to be designed "right the first time" with the right combinations of features that not only satisfy the customers but continually please and delight them by providing increased functionality, comfort, convenience, safety and craftsmanship.

The book is based on my over forty years of experience as a human factors researcher, engineer, manager and teacher, who has performed numerous studies and analyses designed to provide answers to designers, engineers and managers involved in designing car and truck products, primarily for the markets in the United States and Europe. The book is not like many ergonomics textbooks, which compile a lot of information from a large number of references reported in the human factors and ergonomics literature. I have included only the topics and materials that I found to be useful in designing car and truck products and concentrated on the ergonomic issues generally discussed in the automotive design studios and product development teams. The book is really about what an ergonomics engineer should know and do after he or she becomes a member of an automotive product development team and is asked to create an ergonomically superior vehicle.

The book begins with the definitions and goals of ergonomics, historical background and ergonomics approaches. It covers important human characteristics, capabilities and limitations considered in vehicle design in key areas such as anthropometry, biomechanics and human information processing. Next, the reader is led in understanding how the driver and the occupants are positioned in the vehicle space and package drawings and/or CAD models are created from key vehicle dimensions used in the automobile industry. Various design tools used in the industry for occupant packaging, driver vision and applications of other psychophysical methods are described. The book covers important driver information processing concepts and models and driver error categories to understand key considerations and principles used in designing controls, displays, and their usages including current issues related to driver workload and driver distractions.

A vehicle's interior dimensions are related to its exterior dimensions in terms of the required fields of view from the driver's eye points through various window openings and other in-direct vision devices (e.g., mirrors, cameras). Various field of view measurement and analysis techniques and visibility requirements and design areas such as windshield wiper zones, obscurations caused by the pillars, and the required in-direct fields of views are described along with many trade-off considerations. Human factors considerations and night visibility issues are presented to understand basics of headlamp beam pattern design and signal lighting performance, and their photometric requirements.

Other customer/user concerns and comfort issues related to entering and exiting from the vehicle, seating, loading and unloading cargo and other service related issues (in engine and trunk compartment, refueling the vehicle, etc.) are covered. They provide insights into user considerations in designing vehicle body and mechanical packaging in terms of important vehicle dimensions related to body/door openings, roof, rocker panels, and clearances for the user's hands, legs, feet, torso and head, and so forth.

A chapter on craftsmanship covers a relatively new technical and increasingly important area for ergonomics engineers. The whole idea behind the craftsmanship is that the vehicle should be designed and built such that the customers will perceive the vehicle to be built with a lot of attention-to-details by craftsmen who apply their skills to enhance the pleasing perceptual characteristics of the product related to its appearance, touch feel, sounds and ease during operations. Several examples of research studies on measurement of craftsmanship and relating product perception measures to physical characteristics of interior materials are presented.

In addition for the researchers, the second part of the book includes chapters on driver behavioral and performance measurement, vehicle evaluation methods, modeling of driver vision – which illustrates how the target detection distances and legibility of displays can be predicted to evaluate vehicle lighting and display systems, and driver workload to evaluate in-vehicle devices. Discussions on ergonomic issues for the development of new technological features in areas such as telematics, night vision and other driver safety and comfort related devices are included. The second part of the book also presents data and discusses many unique issues associated with designing for different populations segments, such as older drivers, women drivers, drivers in different geographic parts of the world. Finally, the last chapter is focused on various key issues related to future research needs in various specialized areas of ergonomics as well as vehicle systems, and on implementation of available ergonomic design guidelines and tools at different stages of the automotive product design process.

The book can be used to form the basis of two courses in Vehicle Ergonomics. The first course would cover the basic ergonomic considerations needed in designing and evaluating vehicles that are included in Part I – the first eleven chapters of this book. The remaining chapters covered in Part II can be used for an advanced and more research-oriented course.

Vivek D. Bhise
December 26, 2010

MATERIAL FOR WEBSITE

Volume 2:
Visibility Prediction Model – Volume 2 chapter 14
Discomfort Glare and Dimming Request Model – Volume 2 chapter 14
Financial Model of Product Development Program (Revenues and costs) – Volume 2 Chapter 23

About the Author

Vivek D. Bhise is currently a LEO lecturer and professor in post-retirement of Industrial and Manufacturing Systems Engineering at the University of Michigan-Dearborn. He received his B.Tech. in Mechanical Engineering (1965) from the Indian Institute of Technology, Bombay, India, M.S. in Industrial Engineering (1966) from the University of California, Berkeley, California and Ph.D. in Industrial and Systems Engineering (1971) from the Ohio State University, Columbus, Ohio.

From 1973 to 2001, he held a number of management and research positions at the Ford Motor Company in Dearborn, Michigan. He was the manager of Consumer Ergonomics Strategy and Technology within the Corporate Quality Office and the manager of the Human Factors Engineering and Ergonomics in the Corporate Design of the Ford Motor Company where he was responsible for the ergonomics attribute in the design of car and truck products.

Dr. Bhise has taught graduate courses in Vehicle Ergonomics, Vehicle Package Engineering, Automotive Systems Engineering, Human Factors Engineering, Total Quality Management and Six Sigma, Product Design and Evaluations, and Safety Engineering over the past 43 years (1980–2001 as an Adjunct Professor, and 2001–2009 as a professor, and 2009-present as a lecturer) at the University of Michigan-Dearborn. He also worked on a number of research projects on human factors with late Prof. Thomas Rockwell at the Driving Research Laboratory at the Ohio State University (1968–1973). His publications include over 100 technical papers in the design and evaluation of automotive interiors, vehicle lighting systems, field of view from vehicles, and modeling of human performance in different driver/user tasks.

He received the Human Factors Society's A. R. Lauer Award for Outstanding Contributions to the Understanding of Driver Behavior in 1987. He has served on many committees of the Society of Automotive Engineers, Vehicle Manufacturers Association, Human Factors Society, and Transportation Research Board of the National Academies. He is a member of the Human Factors and Ergonomics Society, the Society of Automotive Engineers, Inc. and the Alpha Pi Mu. He has published the following books:

Bhise, Vivek D. 2011. *Ergonomics in the Automotive Design Process.* ISBN: 9781439842102. Boca Raton, FL: CRC Press. (Also translated in Chinese language and published by China Machine Press in China, 2016).
Bhise, Vivek D. 2013. *Designing Complex Products with Systems Engineering Processes and Techniques.* ISBN: 13: 978-1466507036. Boca Raton, FL: CRC Press.
Bhise, Vivek D. 2017. *Automotive Product Development: A Systems Engineering Implementation.* ISBN: 978-1-4987-0681-0. Boca Raton, FL: CRC Press.
Bhise, Vivek D. 2022. *Decision-Making in Energy Systems.* ISBN: 978-0-367-62015-8. Boca Raton, FL: CRC Press.
Bhise, Vivek D. 2023. *Designing Complex Products with Systems Engineering Processes and Techniques.* Second Edition, ISBN: 978-1032203690. Boca Raton, FL: CRC Press.

Acknowledgments

This book is a culmination of my education, experience and interactions with many individuals from the automotive industry, academia and government agencies. While it is impossible for me to thank all the individuals who influenced my career and thinking, I must acknowledge the contributions of the following individuals.

My greatest thanks go to Prof. Thomas H. Rockwell of the Ohio State University. Tom got me interested in human factors engineering and driving research. He was my advisor and mentor during my doctoral program. I learnt many skills on how to conduct research studies, analyze data; and more importantly, he got me introduced to the technical committees of the Transportation Research Board and the Society of Automotive Engineers.

I would like to thank Lyman Forbes, Dave Turner and Bob Himes from the Ford Motor Company. Lyman Forbes, manager of the Human Factors Engineering and Ergonomics (HFEE) Department at the Ford Motor Company in Dearborn, Michigan, spent hours with me discussing various approaches and methods to conduct research studies on various crash-avoidance research issues related to the development of motor vehicle safety standards. Dave Turner from Advanced Design Studios helped anchor ergonomics in the automotive design process and also created an environment to establish a human factors group in Europe. Bob Himes of the Advanced Vehicle Engineering staff helped in incorporating "Ergonomics and Vehicle Packaging" as a "Vehicle Attribute" in the vehicle development process.

The University of Michigan-Dearborn campus provided me with the unique opportunities to develop and teach various courses. Our Automotive Systems Engineering and Engineering Management Programs allowed me to interact with hundreds of students who in-turn implemented many of the techniques taught in our graduate programs in solving problems within many other automotive OEMs and supplier companies. I want to thank Profs. Adnan Aswad, Munna Kachhal and Armen Zakarian for giving me opportunities to develop and teach many courses in the Industrial and Manufacturing Systems Engineering, and Dean Subrata Sengupta for supporting the creation of the Vehicle Ergonomics Laboratory in the new Institute for Advanced Vehicle Systems Building. Roger Schulze, Director of the Institute for Advanced Vehicle Systems got me interested in working on a number of multidisciplinary programs in vehicle design. Together, we developed a number of vehicle concepts such as the Low Mass Vehicle, a new Model "T" concept for Ford's 100th anniversary, and a Reconfigurable Electric Vehicle. We also created a number of design projects by forming teams of our engineering students with students from the College for Creative Studies in Detroit, Michigan. My special thanks also go to James Dowd from Collins and Aikman and the Advanced Cockpit Enablers (ACE) team members for sponsoring a number of research projects on various automotive interior components and creation of a driving simulator to evaluate a number of advanced concepts in vehicle interiors.

Over the past forty-plus years, I was also fortunate to meet and discuss many automotive design issues with members of many committees of the Society of Automotive

Engineers, Inc., the Motor Vehicle Manufacturers Association, the Transportation Research Board and the Human Factors and Ergonomics Society.

I would like to also thank Cindy Carelli from CRC Press, Taylor & Francis Group, for encouragement in preparing the proposal for this book, and Amy Blalock and her production group for turning the manuscript into this book. My thanks also go to Louis Tijerina, Anjan Vincent and Calvin Matle who spent hours reviewing the manuscript and providing valuable suggestions to improve this book.

Finally, I want to thank my wife, Rekha, for her constant encouragement and her patience while I spent many hours working on my computers writing the manuscript and creating figures included in this book.

Vivek D. Bhise
Ann Arbor, Michigan
December 26, 2010

14 Modeling Driver Vision

USE OF DRIVER VISION MODELS IN VEHICLE DESIGN

The objectives of this chapter are to present driver vision models and discuss their applications for assessing driver visibility related problems encountered during the vehicle design process. The problems covered in this chapter are determination of a) visibility of targets under headlamp illumination, b) legibility of displays, and c) veiling glare caused by sun reflections.

The driver vision models considered in this chapter are based on determinations of what the driver can see and read. The models are primarily used to evaluate detection of targets on the roadway due to illumination from vehicle lighting systems, legibility of displays, and evaluation of daytime visibility under situations of veiling glare caused by reflections of the top of the instrument panels into the windshields, and reflections of the top of the package trays into the backlites (rear window glass) while viewing rearwards directly or through the inside rearview mirrors.

This chapter begins with a discussion of the visual contrast thresholds of the human eye. During his research work on driver vision, the author found that the visual contrast threshold curves developed by Blackwell (1952) could be used to predict visibility and legibility under a variety of daytime and nighttime conditions. The prediction capabilities of these models were compared with the visibility distances obtained in seeing distance tests while driving in both the absence and presence of glare caused by the on-coming vehicle headlamps (Bhise, Farber and McMahan, 1977a; Bhise, Farber, Saunby et al., 1977b). The discomfort glare evaluation model developed by DeBoer (1973) was also found to be useful by comparing the predicted values of the DeBoer discomfort index with the dimming request behavior of the drivers on public roads (Bhise, Farber, Saunby et al., 1977b). The visual contrast thresholds were also used to predict the legibility of labels and numerals under day, dawn/dusk and night driving conditions of automotive displays (Bhise and Hammoudeh, 2004). Rockwell, Augsburger, Smith and Freeman (1988) also used a similar model to predict legibility of electronic displays. The veiling glare caused by the reflections of the top of the instrument panels in the windshields were also modeled to predict veiling glare experienced by the drivers and their effects on target visibility (Bhise and Sethumadhavan, 2008a, 2008b).

SYSTEMS CONSIDERATIONS RELATED TO VISIBILITY

Determining the level of visibility of an object is a systems problem because it depends upon characteristics of a number of components of the highway transportation system.

DOI: 10.1201/9781003485605-1

The visibility of an object (a target or a visual detail) depends primarily upon the photometric and geometric characteristics of the object, the visual environment, the vehicle, and the observer's (driver's) eyes and his/her visual information processing capabilities. The important characteristics related to visibility are given below.

Target Characteristics:

a) Location of the target with respect to an observer's eyes and the line of sight (i.e., target distance and angular location of the target with respect to the observer's line of sight)
b) Target orientation (stand-up target or horizontal target on the roadway, e.g., lane marking)
c) Size (i.e., the angular size or the visual angle subtended by the target at the driver's eyes)
d) Reflectance (retro-reflectance, i.e., the ratio of amount of light reflected back into the driver's eyes from the target to amount of light falling on the target from a headlamp)
e) Shape (e.g., length-to-width ratio)
f) Motion/movement of the target
g) Temporal characteristics of the target luminance (e.g., flash rate)
h) Color of the target

Visual Environment Characteristics:

a) Ambient lighting conditions (i.e., illumination from external sources)
b) Road geometry (e.g., lane configurations, curvatures and grades)
c) Background against which a target is seen (i.e., luminance or reflectance of background material)
d) Weather conditions (e.g., transmission and scattering of light through fog, rain, and snow)
e) Glare sources (i.e., their locations with respect to observer's eyes and the line of sight, and luminous intensity directed at the observer's eyes)

Vehicle Characteristics:

a) Driver's (observer's) eye locations in the vehicle (e.g., coordinates of the eyellispse centroids)
b) Vehicle components causing obstructions in the driver's field of view (e.g., pillars, mirrors, and headrests)
c) Headlamp locations and headlamp axes orientations (i.e., coordinates of headlamp centers and horizontal and vertical headlamp mechanical misaim due to body variations, vehicle loading and aerodynamic lift)
d) Headlamp beam patterns (i.e., distribution of luminous intensities of left and right headlamps)
e) Headlamp optical misaim

    f)   Glazing (glass) materials and installation angles (i.e., light transmissivity at installed angle of the vehicle glass through which the target is viewed)

    g)   Graphics in visual displays (i.e., sizes of visual details, their background luminance, visual contrast, colors)

    h)   vehicle location on the roadway and speed

Observer (Driver) Characteristics:

    a)   Observer's age (Note: The visual contrast thresholds, disability glare and discomfort glare increase with an increase in the observer's age.)

    b)   Eye defects (e.g., cataracts and corrections in eyeglasses)

    c)   Seating position and eye locations (with respect to the locations of the SgRP and the eyellipses)

    d)   Eye and head movement behavior (i.e., eye and head-turn angles)

    e)   Psychological and physiological state (e.g., effects of fatigue, alertness/attention/distraction, and substances)

LIGHT MEASUREMENTS

This section provides definitions of light measurement units that are used to compute photometric characteristics of targets and their backgrounds which, in turn, are used to predict visibility and legibility.

LIGHT MEASUREMENT UNITS

Light is defined as visually sensed radiant energy. Radiant energy is the total energy emitted by a source. Only some part of the radiant energy can be sensed by the human visual system and is perceived as "light". Thus, visible light is a part of radiant energy. The light measurement units used to measure the outputs of light sources and resulting luminance of targets and their backgrounds are defined as follows:

1. Luminous energy (Q) or quantity of light is defined as visually sensed radiant energy. It is measured in Talbots (T) with the units Lumen-second (lm s).

2. Luminous flux (Φ) is light power. It is the time rate of flow of light energy, and it is defined as $\Phi = Q/t$ and is measured in Lumen (lm); where t = time in seconds.

3. Luminous intensity (I) is defined as the amount of luminous flux emitted (Φ) in a given solid angle (ω). The unit of luminous intensity is the Candela (cd).

   Thus, $I = \Phi/\omega$, where ω = solid angle measured in steradians (sr). The solid angle is defined as: $\omega = S/r^2$, where S = elemental area (through which the light flux is emitted) on the surface of a sphere of radius r (through which the light flux is emitted).

4. Illuminance or illumination (E) is the density of luminous flux incident upon a surface. It is the quotient of flux (Φ) divided by the projected area A of the illuminated surface when the flux is averaged over the area.

Thus, $E = \Phi/A$. The illumination is measured in the SI metric system in Lux (lx) which has the units of lumens/m^2. The measure of illumination in the English units is Foot-candle (fc), which has the units of lumens/ft^2. The conversion equations are: 1 Foot-candle = 10.76 Lux, and 1 Lux = 0.0929 Foot-candle.

$E = (I/d^2)$ $_*$ cos α, where I = Luminous intensity, d = distance from the source to the receiving plane, and α = the angle of the normal to the plane with respect to the incident light. The inverse square law thus applies here, stating that illumination falling on a surface is directly proportional to the inverse of the square of the distance from its source to the surface.

5. Luminance (L) or photometric brightness is measured in nit or cd/m^2 in the SI metric system, or in footlambert (fL) in the English system. The conversion equations are: 1 nit = 1 cd/m^2 = 0.2919 fL; 1 fL = 3.426 cd/m^2; 1/π cd/ft^2 = 1 fL; where π = 3.142.

For a surface viewed under direct illumination: The luminance of a surface (L) due to illumination (E) incident on the surface (of a target or a background of a target) can be calculated as:

$L = [(\text{Reflectance of the surface}) \times (\text{Illumination})]/\pi = [R \times E]/\pi$
$= R \times E \text{ (fc)} = L \text{ (fL)}$
$= [R \times E \text{ (lux)}]/\pi = L \text{ (cd/m}^2)$

Note: Reflectance (R) is a dimensionless number expressed as a number between 0 to 1 (or in percent 0 to 100%). The reflectance of white matte paint is about 0.85 and the reflectance of matte black paint is about 0.04.

For a surface viewed under transmitted illumination: Luminance of a surface (L) due to illumination (E) incident on the back surface (of a translucent surface of a target or a background of a target) can be calculated as:

$L = [(\text{Transmittance of the surface}) \times (\text{Illumination})]/\pi$
$= [T \times E]/\pi$
$= T \times E \text{ (fc)} = L \text{ (fL)}$
$= [T \times E \text{ (lux)}]/\pi = L \text{ (cd/m}^2)$

Note: Transmittance (T) is a dimensionless number expressed as a number between 0 to 1 (or in percentage 0 to 100%).

6. The total light energy falling on a given material will be transferred as the sum of reflected energy, transmitted energy or absorbed energy. In photometry, we are concerned with the measurement of the reflected or transmitted visible light energy which is sensed by the human eye as luminance.

PHOTOMETRY AND MEASUREMENT INSTRUMENTS

Photometry is defined as the science of measuring visible light based upon the response of an average human observer. A photometer is an instrument used to measure photometric quantities such as luminance, illuminance, luminous flux and luminous intensity. For luminance measurements, a photometer projects an image of a visual detail (e.g., target or background) on a photocell. And for illuminance or illumination measurement it measures light flux falling on a white light-gathering lens placed in front of the photocell. A spectroradiometer is an instrument for measuring the spectral energy radiated by a source. The spectral data can be used to calculate photometric and colorimetric parameters.

VISUAL CONTRAST THRESHOLDS

Blackwell (1952) developed a basic model of the visual detection capabilities of the human eye based on the concept of visual contrast thresholds. A visual target is seen by an observer because of the visual contrast between the target and its background. Visual contrast (C) is defined as follows:

$$C = (|L_t - L_b|)/L_b$$

Where, L_t = Luminance of the target and L_b = Luminance of the background

Figure 14.1 presents three circular targets with different visual contrast values. The two targets on the left have the same luminance of the target (L_t). The middle target is more visible because its background luminance (L_b) is lower than the background of the target on the left. The right target has the same luminance as the background luminance of the left target and its background luminance is the same as the luminance of the middle target. However, the right target is not as visible as the middle target because the luminance and the contrast of the right target are lower.

The contrast (C) must be greater than the threshold contrast (C_{th}) for the target to be visible. The threshold contrast levels were determined by Blackwell (1952) by conducting laboratory experiments. The experiments involved presenting to the

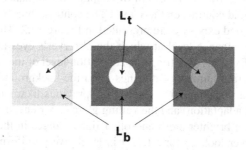

FIGURE 14.1 Illustration of three circular targets with different combinations of target luminance (L_t) and background luminance (L_b).

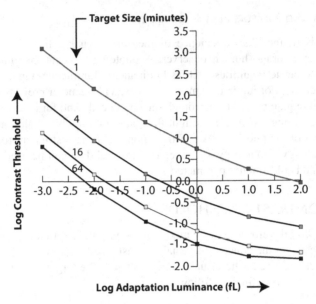

FIGURE 14.2 Visual contrast threshold curves for 1/30th second target exposure used in the target detection model.

subjects a number of circular targets with different diameters and luminance values against backgrounds with different luminance values to create different contrasts levels. Each subject sat in front of a screen and the subject was presented with a different target in each trial. The targets were presented in tachistoscopic exposures ranging from 0.01 s to 1 s. (The tachistoscopic exposures were accomplished by using an electro-mechanical shutter in front of the lens of a slide projector.) The task of the subjects was to determine if a target presented in each trial was visible or not visible. The trials included combinations of target size, background luminance, contrast value and exposure duration. From these data Blackwell developed contrast threshold curves for target sizes ranging from 1 minute to 64 minutes in the plotting space defined by the logarithm of adaptation luminance on the x-axis and logarithm of threshold contrast on the y-axis. The contrast threshold curves obtained under 1/30th of second exposure are presented in Figure 14.2. The adaptation luminance (L_a) is the luminance level at which the observer's eyes are adapted. As the human eye makes fixations over time, the adaptation level of the human eye changes rapidly. The adaptation luminance depends upon the background luminance of the target being viewed, the adaptation level in the previous fixations (due to an effect called the transient adaptation) and the veiling glare experienced by the observer (as scattered light) from brighter areas and other light sources in the observer's visual field. For an observer looking for a target in a relatively uniform luminance background, the adaptation luminance can be assumed to be equal to the luminance of the background.

BLACKWELL CONTRAST THRESHOLD CURVES

The visual contrast threshold data developed originally by Blackwell (1952) and validated for night driving situations by Bhise et al. (1977a) were used to model basic target detection situations in night driving. It should be noted that the contrast threshold curves presented in Figure 14.2 were obtained by fitting Blackwell's threshold data obtained under 1/30th of a second exposure. The log values used in the abscissa and ordinate of the graphs are to the base value of 10. The 1/30th second exposure duration is much smaller than the typical eye fixation durations (of about 1/3rd s) made by the drivers during driving. However, since Bhise et al. (1977a) found that the higher contrast thresholds for 1/30th second (as compared to at 1/3rd s exposure) predicted visibility distances to stand-up targets more accurately under the more difficult actual dynamic driving conditions, the Blackwell's contrast thresholds obtained under 1/30th of second were used here. Overall, the field-observed seeing distances of stand-up and delineation line targets (i.e., painted lane markings) could be predicted within about 13 percent accuracy by the use of the Blackwell thresholds.

Computation of Contrast Values

The target contrast (C) was computed as follows: $C = [|L_t - L_b|] / L_b$

Where, L_t = Target Luminance (fL)
L_b = Background Luminance (fL)
$L_t = R_t \times (E_{lt} + E_{rt}) + L_{at}$
$L_b = R_b \times (E_{lb} + E_{rb}) + L_{ab}$
R_t = Reflectance of the target (fL/fc)
R_b = Reflectance of the background (fL/fc)

(Note: Typical retro-reflectance values of dry pavements range from about 0.03 to 0.10 [Bhise et al., 1977b]; Wet pavement retro-reflection values range from about 0.005 to 0.04 [Bhise et al., 1977b]; Pedestrian summer and winter clothing reflectance range typically from 0.02 to 0.50 and 0.02 to 0.30, respectively [Bhise et al., 1977b].)

E_{lt} = Illumination from the left headlamp falling on the target (fc)
E_{rt} = Illumination from the right headlamp falling on the target (fc)
E_{lb} = Illumination from the left headlamp falling on the target background (fc)
E_{rb} = Illumination from the right headlamp falling on the target background (fc)

The above illumination levels are computed from the luminous intensity (cd) values directed at each target (or its background) point from each headlamp and dividing them by squared values of the corresponding distances from the headlamp to the target/background point.

L_{at} = Ambient luminance of the target (fL)
L_{ab} = Ambient luminance of the background (fL)

(Typical ambient luminance values on rural non-illuminated areas range from about 0.0001 to 0.01 fL, and in urban illuminated areas range from about 0.001 to 0.10 fL (Bhise et al., 1977b))

COMPUTATION OF THRESHOLD CONTRAST AND VISIBILITY DISTANCE

The contrast threshold (C_{th}) was modeled as a function of adaptation luminance (L_a) and target size (θ) in minutes of arc at the eye, and contrast multipliers were used to account for the effects of driver age, confidence in detection judgment, driver alertness level and attention-getting value of the target (conspicuity level). The C_{th} was computed by using the following expression:

$$\text{Log}_{10}\, C_{th} = ((B_0 + 10B_1 L + 100B_2 L^2)/10) + \text{Log}_{10}(T_M)$$

Where

$L = \text{Log}_{10}\,(L_a \text{ in foot-Lamberts})$

$L_a = W_T \times L_b$

$B_0 = 7.4935 - 6.97678\,S + 0.544938\,S^2$

$B_1 = -0.55315 + 0.021675\,S + 0.0003125\,S^2$

$B_2 = 0.007721 + 0.000558\,S + 0.0000175\,S^2$

$S = \text{Log}_2\theta = \text{Logarithm to the base 2 of target size in minutes}$

TS = Target size (feet). It is the diameter of a circle that has the same area as the target.

θ = Target size (minutes) = $[\text{Tan}^{-1}\,(\text{TS/VD})] \times (180.0/\pi) \times 60$

VD = Viewing Distance (feet)

W_T = Windshield transmittance (Typical value is about 0.65 or 65%)

$T_M = M_a \times M_c \times M_{at}$ = Contrast multiplier to account for the effects of observer age, confidence in detection judgment, and level of attention getting characteristic (or conspicuity) of the target. The three multipliers are described below:

   M_a = Multiplier for contrast to account for effects of observer's age (OA) from (Blackwell and Blackwell, 1971; Bhise et al., 1977b)

   = $-0.3796391 + 0.1343982\,OA - 0.0044422\,OA^2 + 0.0000550484\,OA^3$

Figure 14.3 shows a plot of the contrast multiplier (M_a) as a function of the observer's age (OA).

   M_c = Multiplier to account for the observer's confidence (or the probability) in detecting the target. The multipliers were derived from the variability in the contrast threshold reported by Blackwell (1952).

   = 1.00 for 50% confidence

   = 1.58 for 90% confidence

   = 1.78 for 95% confidence

   = 2.24 for 99% confidence

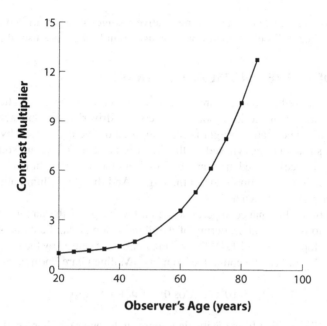

FIGURE 14.3   Contrast multiplier to account for observer's age.

M_{at} = Multiplier to account for attention-getting level (or conspicuity) of the target
    = 1.0 for "just detectable" target
    = 2.5 for "easy to see" target
    = 5.0 for an "unalerted" driver (i.e., to model a driver who is unaware that a target will appear as compared to an alerted driver who is expecting a target to appear in an experimental situation)
    = 10.0 for attention getting

Thus, the total threshold contrast required to see a visual target with the above multipliers should be C_{tha}, which is $C_{th}xT_M$

The target is considered visible if the computed target contrast C is greater than the threshold contrast (i.e., $C > C_{tha}$)

To determine the maximum visibility distance, the above procedure should be iterated by increasing the viewing distance (VD) in steps of 10 to 100 feet until a previously visible target becomes invisible, or by decreasing the viewing distance (VD) in steps of 10 to 100 feet until a previously invisible target becomes visible.

The above contrast multipliers were developed by the author during the experiments and analyses conducted in the headlighting research program at the Ford Motor Company (Bhise et al., 1977b). The approach of using the multipliers is somewhat controversial as the selection of values for the multipliers requires the researcher's judgment and some calibration with actual detection data collected under known field conditions (measured by using a photometer). However, for the purposes of sensitivity analysis and to determine relative changes in visibility with respect to certain reference or baseline conditions, the use of multipliers is a useful approach.

EFFECT OF GLARE ON VISUAL CONTRAST

When one or more light sources are present in a driver's (or observer's) field of view, the illumination from the light sources enters the driver's eyes and gets scattered inside the eyes. The scattered light is superimposed on the image seen by the driver. Figure 14.4 shows a target viewed by the driver in presence of a glare source (called the ith glare source) located at an angle of Φ_i from the line of sight (from the cyclopean eye location of the observer) to the target. And the glare illumination entering the observer's eye is shown as E_i.

The additional luminance superimposed on the image of the target and its background due to the internal scattering of the illumination E_i can be computed using a formula developed by Fry (1954). The veiling luminance caused by the ith glare source is defined as L_{vi} and can be computed from Fry's Veiling Glare Formula as follows:

$$L_{vi} = 10\ \pi\ (E_i\ \text{Cos}\ \Phi_i)/[(\Phi_i + 1.5) \times \Phi_i]$$

Where, E_i = Illumination from the ith glare source in fc, and Φ_i = Glare angle in degrees

If there is more than one glare source, the formula for the total veiling luminance (L_v) obtained by summing the veiling luminance from each of the glare sources can be stated as follows:

$$L_v = 10\ \pi\ \Sigma\ \{(E_i\ \text{Cos}\ \Phi_i) / [(\Phi_i + 1.5) \times \Phi_i]\}$$

The observer's adaptation luminance (L_a), due to the superimposed veiling luminance, will increase as shown below.

$$L_a = L_b + L_v$$

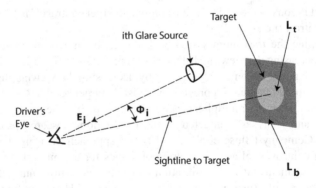

FIGURE 14.4 Target detection situation in presence of a glare source directing glare illumination into the driver's eye.

The luminance contrast of the target will be modified due to addition of the veiling glare to the luminance of the target and the background as follows:

$$C = (| (L_t + L_v) - (L_b + L_v) |) / (L_b + L_v) = (|L_t - L_b|)/(L_b + L_v)$$

The above equation, thus, shows that the in presence of veiling glare, the value of the contrast will always decrease from the value obtained under no glare situation (i.e., without the presence of illumination from a glare source).

STEPS IN COMPUTING VISIBILITY OF A TARGET

Step 1: Measure the distance (D_t) of the target from the driver's eye location

The distance should be measured using the 95th percentile SAE J941 eyellipse. Using the mid-point of the left and right rearmost eyes on the 95th percentile eyellipse will provide the farthest eye location to measure the distance to the target. This distance will cover eye locations of most drivers and provide a conservative estimate of visibility.

Step 2: Determine the projected target size

The projected area of the target (A_p) at the driver's eye is equal to A Cos α.
Where, α = angle between normal to the target surface and the sightline from the selected eye point to the target and A = target area.
Determine target size (TS) = Diameter of a circular target of area A_p
$$= (4A_p/\pi)^{0.5}$$
Target Size (θ) in minutes = $[\tan^{-1}(TS/D_t)](180 \times 60)/ \pi$

Step 3: Determine target Luminance

The target luminance (L_t) can be measured by using a photometer or calculated by knowing target reflectance and the illumination (E_t) incident on the target as $L_t = [(R_t \times E_t) + L_{at}]$.
(Note: It is assumed here that the illumination (E_t) is from a single source and is normal to the target.) If multiple light sources are illuminating the target then illumination from all light sources falling in the target should be added.

Step 4: Determine Luminance of the Background of the Target

The target background luminance (L_b) can be measured by using a photometer or calculated by knowing background reflectance and illumination incident on the background as $[(R_b \times E_b) + L_{ab}]$.
(Note: It is assumed here that the illumination (E_b) is from a single source. If multiple light sources are illuminating the background then illumination from all light sources falling in the background should be added.)

Step 5: Compute Veiling Luminance

> The veiling luminance (L_v) can be measured directly by using a photometer equipped with a Fry veiling glare lens adaptor or it can be computed by using the formula given above, by knowing the glare angle and glare illumination from each glare source.Step 6: Compute Luminance Contrast and its Logarithm
>
> Luminance contrast $C = (|L_t - L_b|)/(L_b + L_v)$
> Compute logarithm of $C = \text{Log}_{10}C$

Step 7: Compute Logarithm of Adaptation Luminance

> Logarithm of Adaptation luminance = $\text{Log}_{10}(L_b + L_v)$

Step 8: Plot Point in the Blackwell Space

> Plot the point with the co-ordinates ($\text{Log}_{10}(L_b + L_v)$, $\text{Log}_{10}C$) in the Blackwell space (on the plot of Blackwell curves, Figure 14.2)

Step 9: Determine Required Contrast Threshold Curve for the Target of Size (θ)

> For the value of target size θ (calculated in Step 2), using Figure 14.2 estimate (or interpolate the location of) threshold contrast curve corresponding to θ. The threshold contrast curve should be adjusted (shifted up/down) by magnitude of $\text{Log}_{10}(T_M)$. Or compute $\text{Log}_{10} C_{tha}$ by using the formula given above.

Step 10: Determination of Target Visibility

> If the point computed in step 8 is located above the adjusted threshold curve obtained in step 9, then the target is visible (i.e., when $C > C_{tha}$)

EXAMPLE 1: TARGET VISIBILITY WITHOUT GLARE

Determine if a 0.3 m (1 ft) diameter target of 0.07 reflectance placed at 63.4 m (208 ft) from the driver's eyes would be visible against a background with 0.03 reflectance placed at 152 m (500 feet) from the driver's eyes illuminated by a 10,000 cd intensity headlamp placed at 61 m (200 feet) from the target. Assume that the ambient luminance of the target and the background is 0.001fL and the driver is 20 years old.

The visibility model available on the publisher's website (called "Blackwell Model Calculations Jan 27 2010") was used to input the data described in the above problem. The output of the model presented in Table 14.1 below and shows that the driver could see the target because the bottom line of the table shows that the value of C/C_{tha} is greater than 1.0.

TABLE 14.1
Application of the Visibility Model for Example 1

INPUTS		Value	Units
1	Luminous intensity of the source directed at the target =	10000	cd
2	Distance of the target from the light source =	200	feet
3	Distance of the background from the light source	500	feet
4	Reflectance of the target ($0 < R_t < 1$) =	0.07	fL/fc
5	Reflectance of the background ($0 < R_b < 1$) =	0.03	fL/fc
6	Viewing distance (Observer's eyes to target) =	208	feet
7	Target size (diameter) =	1	feet
8	Ambient Luminance of the target =	0.001	fL
9	Ambient Luminance of the background of the target =	0.001	fL
10	Glare Source luminous Intensity =	0	cd
11	Distance of the glare source from the observer's eyes =	400	ft
12	Angle of the glare source from the sightline to the target =	2	Degrees
13	Observer's Age =	20	Years
14	Contrast Multiplier to account for confidence, conspicuity, etc.	1	

OUTPUTS		Value	Units
1	Illumination at the target (E_t) =	0.2500	fc
2	Luminance of the target ($L_t + L_v$) =	0.0185	fL
3	Illumination at the background (E_b) =	0.04	fc
4	Luminance of the background ($L_b + L_v$)) =	0.0022	fL
5	Veiling glare luminance due to the glare source (L_v)	0	fL
5	Contrast of the target against background =	7.4090909091	
6	Log Contrast =	0.8697649236	Log contrast
7	Log background luminance =	-2.6575773192	Log luminance
8	θ = Target size (angle subtended at observer's eye) =	16.5275013681	minutes
9	$S = \text{Log}_2 \, \theta$ = Log of Target Size to the base 2	4.046796729	Log_2 (θ in minutes)
10	$B_0 = 7.4935 - 6.97678 \, S + 0.544938 \, S2$ =	-11.8158985775	Coefficient
11	$B_1 = -0.55315 + 0.021675 \, S + 0.0003125 \, S2$ =	-0.4603180047	Coefficient
12	$B_2 = 0.007721 + 0.000558 \, S + 0.0000175 \, S2$ =	0.0102657024	Coefficient
13	Log C_{th} = Log of Blackwell Threshold Contrast	0.7667783639	
14	Age Multiplier (M_a) =	0.9718321	
15	Log C_{tha} = Log Threshold contrast with age & other multipliers =	0.7543696038	
14	Difference between actual and required log contrast = Log C - Log C_{tha} =	0.1153953198	
15	Ratio of C/Ctha (Target is visible if the ratio is greater than or equal to 1)	1.3043535357	

EXAMPLE 2: TARGET VISIBILITY IN PRESENCE OF A GLARE SOURCE

Determine the target visibility in Example 1 above if a glare source of 2000 cd is located at 122 m (400 feet) from the driver and at 2 degrees from the driver's line of sight to the target.

The output of the model presented in Table 14.2 shows that since the value of C/C_{tha} is less than 1.0, the target will not be visible to the driver.

Comparing values provided in Table 14.2 with the corresponding values in Table 14.1, the following observations can be made:

1) The contrast value (given in line 5 of the above output tables) changed from 7.409 when the glare source was absent to 0.041 when the glare source was present.

2) The driver's adaptation luminance (given in line 4 of the above output tables) changed from 0.0022 fL when the glare source was absent to 0.394 fL when the glare source was present.

3) When target distance was increased in increments of 3m (10 feet) in Example 1 (after iterating the model), the maximum distance at which the target was visible was 66.5 m (218 ft) from the driver.

4) When target distance was decreased in increments of 3 m (10 feet) in Example 2 (after iterating the model), the maximum distance at which the target was visible was 45 m (148 ft) from the driver.

TABLE 14.2
Application of the Visibility Model for Example 2

INPUTS	Value	Units
1 Luminous intensity of the source directed at the target =	10000	cd
2 Distance of the target from the light source =	200	feet
3 Distance of the background from the light source	500	feet
4 Reflectance of the target ($0<R_t<1$) =	0.07	fL/fc
5 Reflectance of the background ($0<R_b<1$) =	0.03	fL/fc
6 Viewing distance (Observer's eyes to target) =	208	feet
7 Target size (diameter) =	1	feet
8 Ambient Luminance of the target =	0.001	fL
9 Ambient Luminance of the background of the target =	0.001	fL
10 Glare Source luminous Intensity =	2000	cd
11 Distance of the glare source from the observer's eyes =	400	ft
12 Angle of the glare source from the sightline to the target =	2	Degrees
13 Observer's Age =	20	Years
14 Contrast Multiplier to account for confidence, conspicuity, etc.	1	

TABLE 14.2 (Continued)
Application of the Visibility Model for Example 2

OUTPUTS	Value	Units
1 Illumination at the target (E_t) =	0.2500	fc
2 Luminance of the target ($L_t + L_v$) =	0.4110106853	fL
3 Illumination at the background (E_b) =	0.04	fc
4 Luminance of the background ($L_b + L_v$)) =	0.3947106853	fL
5 Veiling glare luminance due to the glare source (L_v)	0.3925106853	fL
5 Contrast of the target against background =	0.0412960698	
6 Log Contrast =	-1.384091279	Log contrast
7 Log background luminance =	-0.4037211166	Log luminance
8 θ = Target size (angle subtended at observer's eye) =	16.5275013681	minutes
9 $S = Log_2\ \theta$ = Log of Target Size to the base 2	4.046796729	Log_2 (θ in minutes)
10 $B_0 = 7.4935 - 6.97678\ S + 0.544938\ S^2$ =	-11.8158985775	Coefficient
11 $B_1 = -0.55315 + 0.021675\ S + 0.0003125\ S^2$ =	-0.4603180047	Coefficient
12 $B_2 = 0.007721 + 0.000558\ S + 0.0000175\ S^2$ =	0.0102657024	Coefficient
13 Log Cth = Log of Blackwell Threshold Contrast	-0.9790176145	
14 Age Multiplier (M_a) =	0.9718321	
15 Log C_{tha} = Log Threshold contrast with age & other multipliers =	-0.9914263746	
14 Difference between actual and required log contrast =Log C - Log C_{tha} =	-0.3926649044	
15 Ratio of C/C_{tha} (Target is visible if the ratio is greater than or equal to 1)	0.4048881773	

DISCOMFORT GLARE PREDICTION

The perception of the level of discomfort that a driver will experience due to the presence of glare sources can be quantified by asking the driver to rate discomforting sensation using the 9-point discomfort glare scale developed by DeBoer (1973). The 9-point rating scale for measuring the DeBoer index (W) is given below.

    W = 1 – Unbearable glare
      = 3 – Disturbing glare
      = 5 – Just acceptable glare
      = 7 – Satisfactory glare
      = 9 – Just noticeable glare

It should be noted that the value of the DeBoer index (W) decreases with increasing discomfort.

The value of the discomfort glare rating (W) can be also predicted by using an equation developed by DeBoer (1973). The DeBoer equation to predict discomfort glare index (W) caused by multiple glare sources and measured on the 9-point scale is given below.

$$W = 2 \, Log_{10}(1+269.0966 \, L_a) -2 \, Log_{10}[\Sigma \, E_i \, / \, \Phi_i^{0.46}] -2.1097$$

Were,

L_a = Adaptation luminance (fL)

E_i = Illumination from ith glare source into the observer's eyes (fc)

Φ_i = Glare angle between the observer's line of sight and the line from the observer's mid-eye to the ith glare source (minutes).

The above equation was found to be useful by Bhise et al. (1977b) in evaluating glare from on-coming headlamps. The above index (W) was developed under static conditions, and therefore, the magnitude of sensation of discomfort experienced by the driver under dynamic passing situation will be different. Based on the DeBoer index, Bhise et al. (1977b) developed a model to predict the probability of an oncoming driver making a dimming request (i.e., flashing his high beams to signal the opposing driver to dim down to the low beam). The dimming requests made by the on-coming drivers were measured by using a glare vehicle equipped with a variable intensity headlamp system on a two-lane highway. Bhise et al. (1977b) used the DeBoer index to account for the effect of adaptation luminance and glare illumination and used the on-coming driver's dimming requests as a measure of unacceptable discomfort. The probability that a driver will make a dimming request due to the glare caused by the opposing headlamps can be computed as follows:

$$P_D(t) = \{-7.622 - 0.099t_g^2 + 6.34 \, t_g - 1.056 \, [W(t)]^2\}/100$$

Where,

$P_D(t)$ = Probability that a driver will make a request to an on-coming driver to switch to low beam at time t (in seconds) before passing the on-coming vehicle

= 0 if t ≤ 2.5 s

= 0 if W ≥ 7

t_g = Potential glare exposure (s). It is the total time during which the on-coming driver will be exposed to glare headlamps during a meeting situation.

$W(t)$ = DeBoer discomfort index computed at time t before passing of the on-coming vehicle

An Excel model is available in the publisher's website to solve the above two equations (see file called "Visibility and Discomfort Glare Calculations Jan 31 2023"). The inputs and outputs of the model are illustrated in Table 14.3.

TABLE 14.3
DeBoer Discomfort Glare Index and Dimming Request Probability Model Inputs and Outputs

INPUTS		Value	Units
1	Luminous intensity of source 1 directed at the observer's Eyes =	10,000	cd
2	Luminous intensity of source 2 directed at the observer's Eyes =	10,000	cd
3	Distance of source 1 from the observer's eyes =	500	feet
4	Distance of source 2 from the observer's eyes =	500	feet
5	Angle of source 1 from the observer's line of sight	2	Degrees
6	Angle of source 2 from the observer's line of sight	2	Degrees
7	Adaptation luminance of the observer's eyes =	0.01	fL
8	Potential glare Exposure (tg) =	25	sec

OUTPUTS		Value	Units
1	DeBoer Discomfort Glare Index (W) =	3.1	Index
2	Probability of observer making a dimming request =	0.8	

LEGIBILITY

Legibility can be defined as the ability of users to read or decipher the text, graphics, or symbols of a display. Legibility is measured by determining the maximum distance at which the display can be read by the user or by determining the characteristics of an observer (e.g., age, visual acuity) who can read a given display from a given distance. Legibility depends upon the user's visual ability to resolve and discriminate key (or critical) visual details required for acquisition of the displayed information. Thus, legibility assumes some level of processing of information from the display after the image of the display is sensed by the observer's eyes.

For the purpose of predicting the legibility of a display, the problem can be simplified into determining if the user can see a key visual detail in a letter, numeral or a graphic character in the visual display. The key element in reading text is generally considered to be the smallest element such as a stroke or a gap between the strokes of a complex letter such as an "E". To recognize the letter "E", the reader needs to visually discriminate between the following five horizontal details (i.e., see each detail separately): 1) the upper horizontal stroke of the E, 2) the gap between the top and the middle horizontal strokes, 3) the middle horizontal stroke, 4) the gap between the middle and bottom horizontal stokes, and 5) the bottom horizontal stroke. Thus, assuming that the above five details are equal in height (i.e., 1/5th the height of the letter) and the smallest visual detail that a person with normal vision can read in a black letter on a white background in the photopic vision is 1 minute of visual arc subtended at the eye in size, the letter height should subtend at least 5 minutes of visual angle. It should be noted that the width of the vertical stroke of letter E is considered to be same the any of the three horizontal strokes, and thus, its 1 minute width will be discriminable.

FACTORS AFFECTING LEGIBILITY

The legibility of a display will depend upon the characteristics of the display, the driver, the vehicle interior, and the visual environment inside the vehicle. Thus, legibility is a systems consideration, and it should be evaluated by understanding and selecting a proper combination of characteristics of all the components of the system.

The geometric and photometric characteristics of displays that affect legibility are:

Geometric displays characteristics:

a) character (letter, symbol) height (e.g., height of the heated backlite defroster symbol on a push button in a climate control)
b) width of letter or symbol
c) stroke width (or smallest key detail)
d) height-to-width ratio
e) font
f) horizontal spacing between characters or numerals
g) vertical spacing between lines

Photometric display characteristics:

a) wavelength (color)
b) luminance of the background (e.g., luminance of the background of a gauge or screen during daytime, dawn/dust and night)
c) contrast of the visual detail against the background (Note: Contrast can be positive if $(L_t-L_b) > 0$ (i.e., white letter on black background) or negative if (L_t-L_b) (i.e., black letter on white background)
d) luminance variations (i.e., non-uniformity in the luminance of a letter and/or its background due to lighting variations or sunlight and shadows can also create uneven luminance)

Letter/Symbol generation factors:

DOT Matrix

a) number of dots in matrix used to create a detail (e.g., 7 x9 dots used to create a letter)
b) dot element shape
c) dot separation (i.e., spacing between dots)
d) missing, distorted or misplaced dots

Stroke matrix

a) generation technique (e.g., lighted segments vs. script)
b) stroke matrix size (e.g., width of each line segment)

c) writing speed (e.g., speed of moving script)
d) positioning accuracy

The vehicle interior or package characteristics that affect the legibility are:

a) locations of driver's eyes (e.g., eyellipses)
b) location and orientation of the display plane (viewing distances and viewing angles)
c) obstructions of the display caused by the steering wheel, stalks, or other vehicle components.
d) luminance of other interior surfaces close to the display
e) illumination from glare sources falling into the driver's eyes and angular locations of glare sources with respect to the sight lines to the displays (e.g., sunlight reflected from chrome bezel surrounding the display)

The environmental characteristics that affect the legibility are:

a) ambient lighting conditions
b) external illumination (e.g., sunlight and streetlights)
c) adaptation luminance
d) dynamic/transient aspects of lighting conditions (e.g., changes in external lighting, approaching on-coming vehicle headlamps, reflections through mirrors, and chrome/shiny surfaces)

Many research studies on legibility have been reported in the driver vision literature. Cai and Green (2005) reviewed some 112 equations proposed by different authors and found that 22 of them examined automotive displays and screens. They found that the range of letter heights predicted by different equations differed considerably. Only two models that they reviewed were based on visual contrast and adapting luminance. However, none of the models provided the ability to incorporate many variables (required to define the viewing situation and display characteristics) and allowed to input adaptation luminance as a continuous variable, which is possible with the model presented in this chapter.

MODELING LEGIBILITY

As described earlier there are many variables that affect legibility. Many of the variables that affect legibility are the same variables that affect visibility of targets. Therefore, the legibility of a display can be predicted by using the visibility model described earlier. The basic assumption in predicting legibility is that the researcher (or the model user) is able to determine the key visual detail that must be visible to the observer. If that key detail is visible then it is further assumed that the observer will be able to read the content of the display. Thus, to read letters, numerals or symbols we can determine if the smallest visual detail such as the width of a stroke (in a letter or a symbol) can be visible.

A version of the visibility model described earlier in this chapter was created by Bhise and Hammodeh (2004) to predict legibility of displays. The visual contrast threshold data originally published by Blackwell (1952) (described earlier) and later validated for legibility predictions by Rockwell et al. (1988) were used to model the basic contrast thresholds as functions of adapting luminance and target size. For legibility computations, the model provides options to evaluate externally illuminated or back-lighted displays. The user is provided three options to input illumination data by inputting: a) light source intensity, distance from the light source to target (letters, numerals or symbols on the display plane) in English or metric units, b) illumination directed at the target and background and their reflectances, or c) luminances of target (i.e., numerals or letters) and backgrounds. Depending upon the above selected method of data entries, the program shows open data entry boxes (or grayed out the boxes when not needed) for required reflectance or transmittance values. The program also requires other inputs such as letter height, letter height-to-stroke-width ratio, viewing distance, observer age, and level of confidence and determines if the letter or numeral can be read by the observer.

The legibility computation program developed by Bhise and Hommoudeh (2004) was exercised using the following inputs for a backlit display: a) Background luminance = 1 cd/m², b) letter luminance = 3, 5 and 9 cd/m² to obtain contrast of the letters with the background = 2, 4 and 8, respectively, c) letter height = 2 to 12.5 mm, d) letter height-to-stroke-width ratio = 5.0, d) viewing distance = 900 mm, e) ease of reading = 95% confidence, and f) driver age from 20 to 80 years in increments of 10 years. Figure 14.5 presents the relationship between letter height (mm) and driver

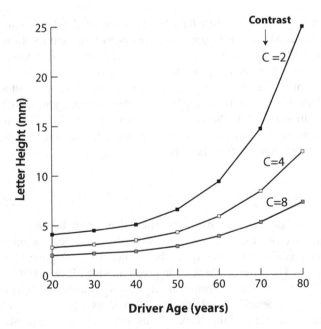

FIGURE 14.5 Illustration of outputs of the legibility model showing letter height required as functions of driver age and contrast (C) of the letters with their background.

age for the three contrast levels used in the above exercise by repeated applications of the model. The exercise involved changing the letter height value and determining the smallest letter height that was legible under each combination of input variables provided above. (Note: Most automotive speedometers have major scale numerals [i.e., MPH numerals in the U.S.] printed in about 6 mm high white letters on a black background).

It is important to note that the contrast ratios of displays are affected considerably by the ambient light reflected and scattered inside the displays by their optics/lenses. Thus, the ergonomics engineer must be careful when determining the contrast ratio for inputting in the program. The author found out that many speedometer graphic elements, which had measured values of contrast ratio above 8 in the laboratory (or design studio) environment, had much lower contrast ratios of about 1.5 to 3 when measured outdoors in the daytime with a photometer. The reduction in contrast due to sunlight falling on the display surface and scattering inside the display elements is also a notable problem with many new technology displays (e.g., LCD).

VEILING GLARE CAUSED BY REFLECTION OF THE INSTRUMENT PANEL INTO THE WINDSHIELD

In designing vehicles, especially with more sloping windshields, it is critical that the right combination of windshield slope or rake angle, instrument panel angle and low reflectance materials (on the top side of the instrument panel) is selected to avoid the problem of degrading the driver's forward visibility. The windshield rake angle is defined as the angle of the windshield surface from the vertical Z-axis measured in the longitudinally located vertical plane at the vehicle centerline (Y= 0). The visibility degradation occurs due to the veiling glare caused by the reflection of the sun- illuminated instrument panel at the windshield. The term "veiling glare" is defined here as the light that is reflected or scattered from the vehicle windshield into the driver's eyes. It is called the "veiling glare" because it creates a veil that is superimposed as unwanted luminance in the driver's view, and thus reduces the driver's visibility. When sunlight falls on the windshield, it illuminates the top part of the instrument panel and its reflection in the windshield is seen by the driver as a veil. The factors that affect this visual effect are: a) the windshield angle (defined in SAE J1100 standard as dimension A121-1 "Windshield Slope Angle". Refer to SAE, 2009), b) windshield type or material (it reflectance characteristics), c) the angle between the instrument panel top surface and the windshield, d) the source of incident light, i.e., the sun during the daytime, e) source (or sun) angle, f) source illuminance (i.e., illumination falling on the windshield), g) instrument panel (top surface) material characteristics such as gloss/reflectance, texture, and color, and h) parting line geometry (joints or discontinuities in the top parts of the instrument panel).

Figure 14.6 presents a driving situation in which the effect of veiling glare will be very critical. The situation involves a driver approaching a dimly-lit area such as a tunnel (or a parking structure) and is looking for an object (or a target) in his path inside the tunnel while the sunlight is falling on his windshield. The visibility of the

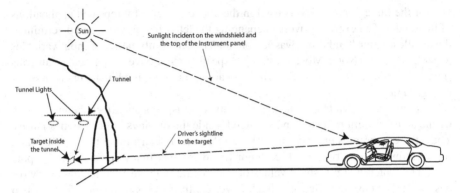

FIGURE 14.6 Veiling glare situation experienced by a driver approaching a target inside a darker tunnel when the sunlight falls on the windshield.

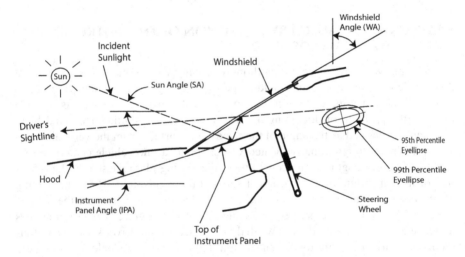

FIGURE 14.7 Ray geometry in the veiling glare situation.

target inside the tunnel will be reduced by the veil created by the reflection of the of the sun illuminated instrument panel into his windshield. The visibility of the target will also depend upon other factors such as the size and reflectivity of the target, its location in the tunnel, tunnel lighting and ambient illumination falling on the target, and the driver's age.

Figure 14.7 presents the ray geometry of the incident sun light as it is reflected into the windshield and viewed by the driver while detecting (observing or looking for) a target. During the early design phases of a vehicle, a vehicle package engineer will generally draw such ray geometry diagrams from various different driver eye points on the eyellipses to assess the windshield and instrument panel architecture for potential veiling glare effects.

A DESIGN TOOL TO EVALUATE VEILING GLARE EFFECTS

Bhise and Sethumadhavan (2008a, 2008b) measured the veiling glare characteristics of windshield reflections and modified the visibility prediction model (described earlier in this chapter) to evaluate the effects of the veiling reflections on the driver's forward visibility. The visibility model can, thus, be used as a design tool to eliminate the distracting and visibility degrading effects of the veiling glare during the early stages of designing a new vehicle.

To develop a model to predict driver visibility under the veil created by reflections in the windhield, Bhise and Sethumadhavan (2008a, 2008b) conducted a two-phase reaserch study. In the first phase, a miniature veiling glare simulator was developed to simulate and measure the veiling glare luminance (using a photometer) on the windshield caused due to reflection of the instrument panel illuminated by simulated sunlight. The measured veiling glare luminance data were used to develop a linear regression model to predict the veiling glare coefficient (VGC) in cd/(m^2 x lux) as a function of the geometric variables asociated with the veiling glare situation and the instrument panel material characteristics. The second phase involved modification of the visibility model to incorporate the veiling glare prediction equation to predict visibility distances to targets in veiling glare situations similar to the "tunnel approaching" situation shown in Figure 14.6.

The veiling glare coefficient (VGC) (defined as the veiling luminance divided by the illumination incident on the windshield) was used as the response measure. A number of stepwise linear regression models to predict the veiling glare coefficient (VGC) as linear and quadratic functions of the following variables were developed: a) the windshield angle (WA), b) instrument panel angle (IPA), c) sun angle squared (SA), and d) gloss value of the instrument panel material (G) [Note: The gloss (reflectance)value was measured by a commercially available gloss meter which measured the value at incident and reflection angles of 60 degrees.], and h) gloss value squared (G2). (Note: The angles WA, IPA and SA are defined in Figure 14.7).

The best-fitting linear model with four variables that was selected to predict VGC is given below:

$$VGC = -0.09276 + 0.00216\ WA + 0.00062\ IPA - 0.0099\ SA + 0.00561\ G$$

The above equation thus shows that the veiling glare effect (or the VGC) increases as the windshield angle, instrument panel angle and the gloss of the instrument panel top are increased. And the veiling effect decreases as the sun angle is increased.

VEILING GLARE PREDICTION MODEL

The situation shown in Figure 14.7 was modeled by using the following variables.

WA = Windshield angle (measured in degrees with respect to the vertical)
IPA = Instrument panel angle (measured in degrees with respect to the horizontal)
SA = Sun angle (measured in degrees with respect to the horizontal)
TS = Target size (diameter in feet of a circle having the same area as the target)

TR = Target reflectance (value between 0 to 1)

BR = Reflectance of target background (value between 0 to 1)

WT = Windshield transmission (Value between 0 to 1)

SI = Sun (i.e., source) illumination incident on the windshield (measured in lux in the plane normal to the sun rays)

TI = Tunnel illumination (measured in lux in the horizontal direction)

C = Target contrast

$$= [(TR \times TI) - (BR \times TI)] / [(BR \times TI) + (VGC \times SI)]$$

(Note: All basic calculations in the model were conducted using English units. The illumination and VGC input values were in metric units.)

θ = Target size in minutes = $[Tan^{-1} (TS/VD)] \times (180.0/\pi) \times 60$

TS = Target size in feet. (It is the diameter of a circle that has the same area as the target).

VD = Viewing Distance to the target in feet

The visual contrast threshold data developed originally by Blackwell (1952) were used to model basic target detection using the same equations provided earlier.

The target was considered visible if the computed target contrast C was greater than the threshold contrast (i.e., $C > C_{th}$) and the value of the answer parameter (ANS) was set equal to 1. Thus, ANS = 1 if the target is visible (detectable); and 0 if not visible.

The visibility distance was determined by iterating the model by using different values of the viewing distance (VD). To compute the visibility distance, the viewing distance (VD) value was first set low and then it was incremented by 3 or 15.2 m (10 or 50 feet) in each iteration until the farthest distance beyond which the value of ANS became "0".

The visibility prediction model was programmed using the Microsoft Excel application. The interface screen showing the inputs and outputs of the model is presented in Table 14.4.

MODEL APPLICATIONS ILLUSTRATING EFFECTS OF DRIVER AGE, SUN ILLUMINATION AND VEHICLE DESIGN PARAMETERS

Baseline Situation

To illustrate the capabilities of the model in predicting visibility distances to a target located in the tunnel as the driver approaches the tunnel with sunlight falling on his/her vehicle windshield, a baseline situation was created. The primary parameters of the baseline situation are shown in rows numbered 1 to 15 of Table 14.4. Thus, the baseline situation can be described as: a 45-year-old driver in a vehicle with 64 degree windshield angle, 0 deg. instrument panel angle and 1.8% gloss, 10,000 lux illumination falling on the vehicle windshield at 20 deg sun angle, approaching a

TABLE 14.4
Model Input Variables and Outputs Showing the Values of the Baseline Situation

Sr. No.	Variable	Inputs for Daytime Driving Condition	Baseline
1	SI	Sun Illumination on windshield (lux)	10000
2	SA	Sun Angle from horizontal (degrees)	20
3	TI	Illumination on the target (lux)	5000
4	TR	Reflectance of the target (0<TR<1)	0.1
5	BR	Reflectance of the background (0<BR<1)	0.05
6	VD	Viewing distance (feet)	1600
7	TS	Target size in equivalent diameter (feet)	2
8	OA	Observer age (years) =	45
9	CME	Contrast Multiplier for confidence or ease	1
10	AC	Adjustment to Contrast Thresholds (in Log C)	0
11	IPA	Instrument panel top angle from horizon (degrees)	0
12	G	Material Gloss (0.8 to 2.65)	1.8
13	WA	Windshield angle from vertical (degrees)	64
14	WT	Windshield transmission along normal to the glass	0.7
15	WR	Windshield reflectance at 20 deg (considered in VGC)	1
16		**Outputs**	
17	VGC	Veiling Glare Coefficient of Instrument Panel Material (Lv in cd/(sq.m x lux))	0.035778
18	LT	Target luminance (fL)	74.2237343476
19	LB	Background luminance (fL)	37.1118671738
20	LV	Veiling luminance (fL)	104.435982
21	LA	Adapting luminance (fL) = LB +LV	141.5478491738
22	CA	Contrast of the target against adapting luminance	0.2621860197
23	LCA	Log contrast	-0.5813904695
24	LLA	Log adapting luminance	2.1509032746
25	TA	Target subtended angle (minutes) at the observer's eye	4.2971812254
26	LLAU	Log adapting luminance upper bound	2
27		Contrast Multipliers:	
28	AM	Age Multiplier	1.68911035
29	CM	Confidence Multiplier	2
30	EM	Exposure Multiplier	1
31	TCM	Total Contrast Multiplier	3.3782207
32		Blackwell Model Calculations:	
33		Theta	2.1033906213
34		B0	-4.7704505248
35		B1	-0.5061764295
36		B2	0.0089721164
37	LTHC	Log Threshold Contrast with above total multiplier =	-0.6018252376
38	THC	Threshold Contrast with above total multiplier =	0.2501351716
39		**RESULT**	
40		Detectable at 1/30th sec Blackwell Threshold Contrast	
41	ANS	with the above total multiplier (1 = Yes; 0 = No) =	1

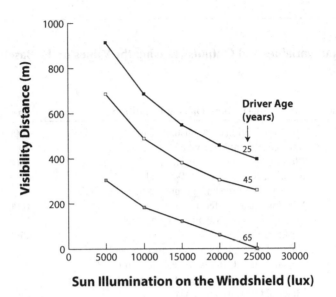

FIGURE 14.8   Effect of sun illumination and driver age on visibility distance to a 0.61 m (2 ft) diameter target under the baseline situation.

0.61m (2 feet) diameter target with 10% reflectance placed in the tunnel. The target is illuminated with 5000 lux (from tunnel and ambient lighting) and the background of the target is assumed to have 5% reflectance. In this baseline situation, the predicted visibility distance of the target was 488 m (1600 feet).

Effect of Incident Sunlight and Driver Age

The effect of sunlight illumination incident on the windshield was evaluated by conducting four additional prediction runs under the above baseline situation by changing the sun illumination from 10,000 lux to 5,000, 15,000, 20,000 and 25,000 lux and using driver ages of 25, 45 and 65 years. The maximum viewing distances at which the target was visible (i.e., ANS=1, as the viewing distances were incremented in steps of 15.2 m (50 feet)) are shown in Figure 14.8 as the visibility distances to the target. The visibility distances decreased as the sun illumination level increased (due to increased veiling luminance). The visibility distances also decreased as the driver's age increased due to an increase in the contrast thresholds with an increase in the driver's age. The visibility distances of the 65-year-old driver were about 457 m (1500 feet) shorter than the 25-year-old driver under the same situation (i.e., the separation between the visibility distance curves of 25- and 65-year-old drivers in Figure 14.8). The veil at 25,000 lux reduced the visibility for a 65-year-old driver to "zero" distance.

Effect of Windshield Angle and Sun Angle

Figure 14.9 shows the effect of the windshield angle and the sun angle on the baseline situation (10,000 lux sun illumination) as the windshield angle was increased from 55

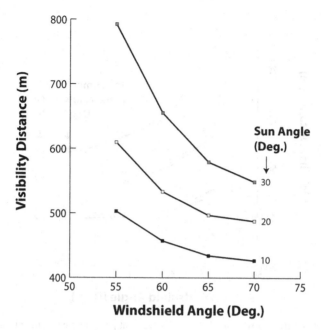

FIGURE 14.9 Effect of windshield angle and sun angle on visibility distance to a 0.61 m (2 ft) diameter target under the baseline situation.

to 70 degrees from the vertical and the sun angle was increased from 10 to 30 degrees above the horizontal. The luminance of the veil increased with an increase in the windshield angle and with a decrease in the sun angle, which in turn reduced the visibility distance. Thus, the worst visibility distance was obtained when the windshield rake angle was 70 degrees, and the sun angle was at 10 degrees.

Effect of Variables Affecting Vehicle Design

The design and appearance of the vehicle can be directly affected by the windshield angle, the instrument panel angle and the material on the top of the instrument panel. Therefore, a sensitivity analysis on visibility distance was conducted by using the above three variables. The results of the analysis are presented in Figure 14.10. The dashed and solid-lined curves in this figure are for gloss values of 2.7 and 0.8, respectively. The results are especially useful in the considerations of trade-offs between the interior and exterior design variables of the vehicle. For introducing higher raked windshields, the instrument panel angle can be decreased and/or the top of the instrument panels can be made (or covered) with materials with lower gloss values. For example, Figure 14.10 shows that visibility with "0 deg. instrument panel angle at 0.8 gloss value" will be similar to "20 deg. instrument panel angle at 2.7 gloss value". Thus, if materials of higher gloss values are selected for the top of the instrument panel, the top surface of the instrument panel should be sloped down more (i.e., away from the windshield).

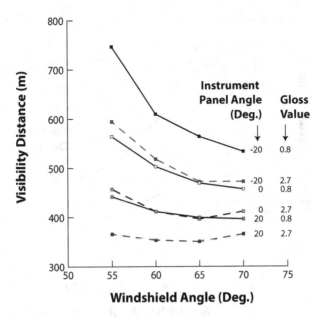

FIGURE 14.10 · Effect of windshield angle, instrument panel angle and the gloss value of the instrument panel material.

CONCLUDING REMARKS

The models presented in this chapter were found by the author to be very useful in teaching concepts and understanding variables related to visibility and legibility to students enrolled in the Automotive Systems Engineering program. The models presented in this chapter can be downloaded from the publisher's website. The reader is encouraged to exercise the models to gain a better understanding of the sensitivity of different photometric, geometric and observer related factors. The models are not only useful as educational tools, but they can also serve as guides in evaluating trade-offs between many variables related to the vehicle exterior and interior designs.

REFERENCES

Bhise, V. and S. Sethumadhavan. 2008a. Effect of Windshield Glare on Driver Visibility. *Transportation Research Record (TRR), Journal of the Transportation Research Board*, No. 2056, Washington, D.C.

Bhise, V. and S. Sethumadhavan. 2008b. Predicting Effects of Veiling Glare Caused by Instrument Panel Reflections in the Windshields. SAE paper no. 2008-01-0666. *International Journal of Passenger Cars – Electronics Electrical Systems*, 1(1): 275–281. Society of Automotive Engineers, Inc., Warrendale, PA.

Bhise, V. D. 2007. Effects of Veiling Glare on Automotive Displays. Presented at the Society of Information Display Vehicle and Photons Symposium held in Dearborn, Michigan.

Bhise, V. D. and C. C. Matle. 1985. *Review of Driver Discomfort Glare Models in Evaluating Automotive Lighting*. Presented at the 1985 SAE International Congress, Detroit, Michigan.

Bhise. V. D. and C. C. Matle. 1989. Effects of Headlamp Aim and Aiming Variability on Visual Performance in Night Driving. *Transportation Research Record*, No. 1247, Transportation Research Board, Washington, D.C.

Bhise, V. D. and R. Hammoudeh. 2004. A PC Based Model for Prediction of Visibility and Legibility for a Human Factors Engineer's Tool Box. *Proceedings of the Human Factors and Ergonomics Society 48th Annual Meeting*, New Orleans, Louisiana.

Bhise, V. D., E. I. Farber and P. B. McMahan. 1977a. Predicting Target Detection Distance with Headlights. *Transportation Research Record*, No. 611, Transportation Research Board, Washington, D.C.

Bhise, V. D., E. I. Farber, C. S. Saunby, J. B. Walnus and G. M. Troell. 1977b. Modeling Vision with Headlights in a Systems Context. SAE Paper No. 770238, 54 pp. Presented at the 1977 SAE International Automotive Engineering Congress, Detroit, Michigan.

Bhise, V. D., C. C. Matle, and E. I. Farber. 1988. Predicting Effects of Driver Age on Visual Performance in Night Driving. SAE paper no. 881755 (also no. 890873), Presented at the 1988 SAE Passenger Car Meeting, Dearborn, Michigan.

Blackwell, H. R. 1952. Brightness Discrimination Data for the Specification of Quantity of Illumination. *Illuminating Engineering*, XLVII(11): 602–609.

Blackwell, O. M. and R. H. Blackwell. 1971. Visual Performance Data for 156 Normal Observers of Various Ages. *Journal of Illuminating Engineering Society*, 1(1), 3–13.

Cai, H. and P. Green. 2005. Range of Character Heights for Vehicle Displays as Predicted by 22 Equations. *Proceedings of the 2005 SID Vehicle Display Symposium*, Dearborn, Michigan.

DeBoer, J. B. 1973. *Quality Criteria for the Passing Beam of Motorcar Headlights*. Paper Presented at the CTB Meeting in Walldorf, Germany.

Fry, G. A. 1954. Evaluating Disabling Effects of Approaching Automobile Headlights. *Highway Research Board Bulletin*, No. 89.

Hoffmeister, D. H. and V. D. Bhise. 1978. A Driver Glare-Discomfort Model to Evaluate Automotive Stop Lamp Brightness. *Proceedings of the 1978 Annual Meeting of the Human Factors Society*, Detroit, Michigan.

Rockwell, T. H., A. Augsburger, S. W. Smith and S. Freeman. 1988. The Older Driver – A Challenge to the Design of Automotive Electronic Displays. *Proceedings of the Human Factors Society – 32nd Annual Meeting*.

Society of Automotive Engineer, Inc. 2009. *The SAE Handbook*. Warrendale, PA: Society of Automotive Engineer, Inc.

15 Driver Performance Measurements

INTRODUCTION TO DRIVER MEASUREMENTS

To evaluate different vehicle designs, vehicle features, and the effects of changes or improvements made in vehicle designs, an ergonomics engineer should be able to measure and demonstrate how well the driver performs in different tasks while using the vehicle. Currently, there are no standardized measures (or variables) or methods for measurement of driver behavior and driver performance. A number of researchers have used many different measures, and there is some agreement on the general approaches for performance measurements. However, many differences in defining even commonly used measures – such as task completion times, variability in lane position and errors – exist due to differences in objectives of the studies, instrumentation, experimental procedures and data collection techniques. Thus, the measurement problem occurs due to differences and/or inconsistencies between researchers in determining "what to measure" and "how to measure" in any given driving or vehicle use situation.

Therefore, the objectives of this chapter are (a) to review various variables used in measuring driver performance in various tasks involved in vehicle uses; (b) to provide the reader a better understanding into issues and problems associated with the measures; and (c) to develop a background in evaluation of vehicle designs by measuring driver and vehicle outputs.

CHARACTERISTICS OF EFFECTIVE PERFORMANCE MEASURES

For any measure to be acceptable and useful, it should meet certain key characteristics. The following characteristics were based on effective safety performance measures presented by Tarrants (1980).

1. *Administrative feasibility*: The measuring system or measuring instrumentation used to obtain the value of the measure must be practical, that is, one must be able to construct it and use it quickly and easily without excessive costs. Thus, an ergonomics researcher should be able to use the measurement system to make the necessary measurements, and the vehicle development

DOI: 10.1201/9781003485605-2

team should be able to set targets by using the measure and determine if the ergonomic goals have been reached.

2. *Interval Scale*: The measurement system should be able to provide the measure by using at least on an interval scale. The interval scale should be graduated with equal and linear units, that is, the difference between any two successive point values should be the same throughout the scale.

It should be noted that there are four types or orders of measurement scales: nominal, ordinal, interval and ratio scales – with ascending order of power to perform mathematical operations. The nominal scale (the most primitive) is used for categorizing, naming or numbering (e.g., model numbers) of items. One can only analyze the data obtained by using the nominal scale by counting frequencies or percentages of values in each category. The ordinal scale is used to order or rank items. But the distance measured on a scale between ranked items (that is, the difference between their scale values) may not be equal. Thus, in ranked items, one can only conclude that an item with a higher rank is better than another item with a lower rank. The interval scale has equal intervals, but the zero point on the interval scale is arbitrary (e.g., like a temperature scale). On an interval scale, the difference between any two items measured on a scale can be determined by the difference between their two respective scale values. The ratio scale is the most informative (or quantitative) as ratios of quantities defined by the scale values can be constructed. For example, a 10-lb weight is two-times heavier than a 5-lb weight. The ratio scale also contains an absolute zero (i.e.. the point of "no amount"). Thus, we should make sure that the measure we select should use the highest possible order of scale – with the interval scale as the minimum order of acceptable scale.

3. *Quantifiable*: A quantitative measure will allow comparison between any two values in terms of at least a difference on an interval scale. The quantitative measure should permit application of more reliable statistical inference techniques. (Note: A non-quantitative, i.e., a qualitative, measure limits statistical inference [due to use of data on nominal and/or ordinal scales] and opens the way for individual interpretation. For example, if the result of a speedometer comparison study states that "the analog speedometers are better than digital speedometers", the reader does not have sufficient information on the magnitude of improvement gained by the use of an analog speedometer.)

4. *Sensitivity*: The measurement technique should be sensitive enough to detect changes in a product characteristic of the product or user performance to serve as a criterion for evaluation. (Note: A tiny diamond cannot be measured on a cattle scale.)

5. *Reliability*: The measurement technique should be reliable, that is, it should be capable of providing the same results for successive applications in the same situations.

6. *Stability*: If a process does not change, the performance level obtained from the measure at any another time should remain unchanged.

7. *Validity*: The measure should produce information that is representative of what is to be measured. This is particularly important because in ergonomic research many different types of measures in a wide range, from indirect and surrogate measures to direct measures, can be used. This issue is discussed later in this chapter.

8. *Error-free results*: An ideal measuring instrument should yield results that are free from errors. However, in general, any measurement will have some constant and random errors. The ergonomics engineer needs to understand the sources of such errors and minimize them. The errors can also be statistically isolated and estimated by their sources (or effects of the sources).

DRIVING AND NON-DRIVING TASKS

It is important to understand that the driver performs a number of tasks while driving. These tasks can be classified as follows:

1. *Lateral Control of the Vehicle*: Maintaining lateral control (left-right movements) of the vehicle within a given driving lane and between lanes in multilane roadways.

2. *Longitudinal Control*: Maintaining longitudinal (fore-aft) control of the vehicle on the roadway and maintaining a safe distance from the lead vehicle in the same lane and other vehicles during maneuvers such as passing, lane changes and lane merges.

3. *Roadway Monitoring*: Acquiring information about the state of the roadway and traffic (e.g., pavement surface characteristics; viewing signs, signals and traffic control devices; detecting/monitoring other vehicles and objects on, or on the sides of, the roadway; using inside and outside mirrors to view traffic in adjacent lanes).

4. *Crash Avoidance*: Avoiding a collision with objects in the vehicle path.

5. *Route Guidance*: Obtaining information from route guidance signs or memorized roadway landmarks to follow a route to the intended destination.

6. *Using In-Vehicle Controls and Displays*: Acquiring information from displays and control settings to understand vehicle state and operating controls.

7. *Other Non-Driving Tasks*: Performing tasks that are not required to drive the vehicle, for example, conversing with passengers, using car phones or other devices, eating, reading materials, and searching/grasping objects (e.g., cups, coins, maps, and papers)

The driver's actions and performance in each of the above mentioned tasks can be measured as functions of time and/or distance, and occurrence frequencies (or rates) of many predefined events can also be measured. The measurements can provide information on a) state of the driver, b) driver outputs, c) driver behavior, d) driver performance, e) driver preferences, f) driver encountered problems and difficulties, g) state of the vehicle, and e) vehicle motion with respect to the highway.

DETERMINING WHAT TO MEASURE

The problem of "what to measure" depends on the researcher's understanding of the driver's tasks and the researcher's motivation and objectives in conducting the measurements. When a driver takes a vehicle out for a trip, the researcher assumes that the driver has certain objectives in making the trip under the driving situation. Figure 15.1 presents a flow diagram to help understand links between the driver's objectives and different types of measures that can be used to evaluate his driving performance.

The driver's objectives that are related to how he or she wants to make the trip depend upon the driver's understanding of his/her capabilities, desires, characteristics of his/her vehicle and the driving situation. For example, the driver may be late for a job interview and would like to make the trip of 50 km within 0.5 h while also making sure that he/she drives safely near the speed limit. The driver's ability to make the trip depends upon his/her characteristics and capabilities as well as the characteristics of his/her vehicle (see top boxes in Figure 15.1). The driver-vehicle combination is then driven on the trip route which has its characteristics such as road geometry, traffic conditions, pavement surface, and so forth. Thus, the driver behavior (i.e., how he/she will act, do or behave during the trip) will depend upon the characteristics of the driver, the vehicle, the roadway and the driving situation. While driving on the route, the driver gets his visual information from the fields of view available from the

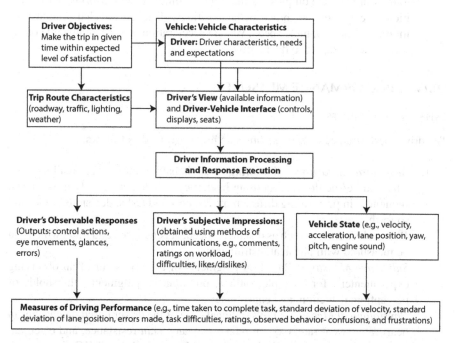

FIGURE 15.1 Flow diagram illustrating links between driver's objectives and driving performance measures.

vehicle, interior displays and operates controls to follow the roadway (see Driver's View box in Figure 15.1).

To measure how well the driver is performing his driving tasks during this trip, we can measure the following (see lower half of Figure 15.1):

1) *Driver's Observable Responses*: We can observe driver's responses such his visual information acquisition behavior through measurements of eye movements, head movements, eye glances, time spent in viewing different objects, and his control movements from measurements of hand and foot movements while operating the steering wheel and the pedals. We can also measure the physiological state of the driver by measuring the driver's heart rate, sweat rate, and so forth.

2) *Driver's Subjective Responses*: We can also develop a structured question-naire and ask the driver a number of questions at different points in the route (if an experimenter is present) or at the end of the trip to understand his driving problems, difficulties, confusions frustrations, and situational awareness issues; and also ask him/her to provide ratings (e.g., using scales, see Chapters 16 and 19) on his workload, comfort, and ease in using different controls and displays.

3) *Vehicle State*: We can also record the state of his vehicle by installing meas-uring instruments (e.g., sensors with data acquisition systems) in the vehicle to measure vehicle outputs (as functions of time) such as steering wheel pos-ition, accelerator and brake pedal positions, distance traveled, lateral pos-ition in the lane, vehicle speed, vehicle acceleration, and heading angle of the vehicle with respect to the roadway.

DRIVER PERFORMANCE MEASURES

TYPES AND CATEGORIES

The driver performance measures, thus, can be categorized as follows:

1. *Behavioral Measures*: Measuring what behavior the driver exhibits (i.e., what did he/she do) by recording his/her eye movements, body movements, sequences in performing different movements and tasks, decisions made, and so forth.

2. *Physical Measures*: such as distance, speed, acceleration, and time, which can be measured with physical instruments.

3. *Subjective Measures*: Based on judgments of the driver (or of an observing experimenter), for example, ratings, preferences, judgments, thresholds of perception, detection, and equivalency.

4. *Physiological Measures*: Physiological state of the driver based on changes in heart rate, sweat rate, oxygen intake, galvanic skin resistance, and electrical activities in different skeletal muscles (EMG) or the heart (EKG).

5. *Accident-based Safety Performance Measures*: Number of accidents of a given type (rear-end, head-on, run-off-the road), accident rate (number of

accidents per million km travelled), and accident severity (property damage, injury level).

6. *Equivalency-based Measures*: Comparisons of performance or driver judgments under two different conditions to determine if they are equal (or the same as) or different.

7. *Monitory Measures*: Measures based on costs such as trip costs, energy consumption, costs and benefits related to the trip (e.g., driver's willingness to pay for an outcome).

The ultimate measure of driver performance from the safety viewpoint is based on the occurrences of accidents, such as the number of accidents of different types (e.g., ran-off-the-road, front-end, rear-end or side collision) or their accident rates (e.g., number of accidents of a given type per 100,000 km or miles driven).

SOME MEASURES USED IN THE LITERATURE

Some commonly used measures of driver performance used in the studies reported in the literature are based on statistics such as mean, median, standard deviation, percentages, and percentile values of the data collected from the following:

1. Velocity.
2. Lane position.
3. Lane departures (lane violations, lane exceedances).
4. Lane changes.
5. Time-to-lane crossing.
6. Steering wheel movements (reversals, rates).
7. Headway (following distance, gap between two vehicles measured in distance or time).
8. Acceleration or deceleration (lateral and longitudinal).
9. Total time spent in looking at a given location or a display.
10. Glance durations while viewing given objects (e.g., speedometer, radio, mirror, sign).
11. Number of glances made to use a device or to complete a given task.
12. Eye fixation durations while viewing a given object.
13. Number of eye fixations made on a given object.
14. Percentage of time eyes were closed.
15. Blink (or eye closure) rate.
16. Detection rates of targets or events.
17. Detection (visibility) distance (of objects on the roadway or roadside).
18. Hand involvement time (i.e., time spent by the driver's hand away from the steering wheel to perform a task).
19. Eye involvement time (i.e., total time spent away from the forward road scene).
20. Driver errors (e.g., lane intrusions, slowed down, excessive speed, misread a display, operated a wrong control, turned control in opposite direction, legibility errors, omission [forgetting] errors, and took a wrong turn).

21. Traffic tickets received in a given period.
22. Task completion time.
23. Reaction time to a given signal (or an event).
24. Brake reaction time (time elapsed after a signal or an event to depress the brake pedal).
25. Accelerator release time (time elapsed after a signal or an event to release the accelerator pedal).
26. Accelerator to brake pedal transition time.
27. Time-to-collision.
28. Accident involvement (e.g., ran-off-the-road, collisions with other vehicle(s), fixed objects, pedestrians, animals). Accident frequency, rates (e.g., number of accidents per 100,000 km or miles of travel).

For definitions and discussions on differences between many of the above measures used by different researchers, an interested reader should refer to Savino (2009).

RANGE OF DRIVING PERFORMANCE MEASURES

The range of driver performance measures that can be used extends from the measurement of some early events, actions or steps in a task to the measurement of final outcomes. For example, the range of events that can be measured in a target detection task while driving can involve measurements of: locations and durations of eye fixations (eye search patterns), target detection response (correct detection or failure), reaction time to detect the target, detection distance, lane position variability while searching for the target, steering wheel position, erratic or evasive maneuver as a result of late or no detection, and accident (if it occurred) resulting from non-detection of the target.

In addition, behavioral measures can provide information on how, what and when the driver performed certain predefined steps. Whereas the measures such as total time spent, types and numbers of errors committed, percentage of times the task was completed in an allocated time – all provide information on how well a given task was completed.

The physiological measures can provide information on the state of the driver's body functions (i.e., how the human body is responding while or after performing the task) by measuring variables such as heart rate, electro-myographic (EMG) potentials, sweat rate, and brain waves.

Subjective measures are also very useful where physical measurement instruments are not available and the subject can be asked to describe problems encountered during a task and provide ratings using scales developed to provide impressions of the subject on the task-related characteristics, such as level of difficulty or ease, magnitude of spatial dimensions, workload, and comfort. (Also see Chapter 16 on driver workload measurements).

Thus, in determining what measures to select for a given study, it is important to ensure that the researcher can obtain useful and valid information. The skill in selecting performance measures depends upon the researcher's knowledge of research literature in the problem area, depth of the researcher's human factors research experience, data sensing and recording equipment availability, time and resources available, and the researcher's experience in statistical data analysis.

The selection of the dependent variables, whether they are behavioral, or performance based, will depend upon the problem and the driver's tasks associated with the issues in the problem. The behavioral variables are generally based on observations of driver movements and actions related to the task being performed. The performance measures, on the other hand, measure how well the driver performed the task. Some commonly used driver behavioral measures in operating in-vehicle devices are number of glances made away from the road, glance durations, percentage of time spent on the task, and sequence of button pushes. Some examples of performance measures are variability in lane position, variation in speed, percentage of tasks completed correctly, and number errors made during task completion.

In experimental evaluations, the dependent variables are generally related to (and affected by) characteristics of drivers (e.g., young versus old, males versus females, experienced versus inexperienced, familiarity of the driver, country of origin or nationality of the driver), characteristics of the driving environment (e.g., day versus night, dry versus wet road, traffic speed, traffic density, and straight versus curved road) and the characteristics of the vehicle design (e.g., analog versus digital display, locations and types of controls, different controls and displays layouts, vehicle features or design configurations, and vehicles produced by different manufacturers in a given market segment).

SOME STUDIES ILLUSTRATING DRIVER BEHAVIORAL AND PERFORMANCE MEASUREMENTS

STANDARD DEVIATION OF LATERAL POSITION

The standard deviation of lateral position provides quantitative information on the variability in maintaining lane position while driving. The standard deviation is computed over a number of measurements (samples) of lateral position data, usually sampled at a preset time or distance interval (selected by the researcher) collected during driving on a test road section. Larger values of the standard deviation suggest that the driver had difficulty in driving within the left and right markings defining the driving lane and any intrusions in the adjacent lanes would mean that the driver could have an accident with a vehicle in an adjacent lane or could run off the road. *The Manual of Uniform Traffic Control Devices* (USDOT, 2003) requires that lane width delineated by lane line pavement markings should not be less than 3 m (10 ft). The lane width on the interstate highways is 3.66 m(12 ft) (AASHTO, 2005; Fitzpatrick et al., 2000).

Green and Shah (2004) found that the most commonly reported measure of driving performance was the standard deviation of lane position. They examined the data on standard deviation of lane position reported in 36 studies and found that the standard deviation values ranged between 0.05 to 0.6 m with the mean value of the standard deviation 0.24 m for studies conducted on roads. The mean values of the standard deviation of lane position in studies conducted in the driving simulators and test tracks were 0.30 m and 0.22 m respectively. They also found that the standard deviation of lane position increased slightly (0.002 m/yr) with the driver age.

Lambert, Rollins and Bhise (2005) conducted a study measuring driving performance using a fixed-base driving simulator and found that the standard deviation of lane position increased by 40–100 percent while performing common driver-induced distraction tasks such as reading a message from a text pager, identifying cross streets on a map, and reading a step in written directions, as compared to the average standard deviation of 0.3 m when the drivers were not performing any other tasks other than just driving on the same road. To perform these more demanding distracting tasks, the drivers made more than 3 eye glances away from the forward road scene.

STANDARD DEVIATION OF STEERING WHEEL ANGLE

Many researchers have measured steering wheel angle and used the standard deviation of the steering wheel angle as a measure to study the change in the driver's activity during test situations. Green and Shah (2004) found that the standard deviation of steering wheel angle was one of the most commonly reported measures of driving performance. They examined the data on standard deviation of steering wheel angle reported in seven studies and found that the mean value of the standard deviation was 1.59 degrees.

STANDARD DEVIATION OF VELOCITY

The standard deviation of forward velocity is a measure of a driver's ability to drive at a constant speed. The speed changes occur due to variables related to the driver (e.g., attention and distractions) and changes in the characteristics of the roadways, traffic, weather, and so forth. Any increases in the standard deviation of velocity can lead to increases in accident rates.

VEHICLE SPEED

The amount of information that the driver needs to process to maintain his vehicle within the lane increases with an increase in the vehicle speed. Fitzpatrick et al. (2000) have shown that driving speed is affected by lane width. They found that the speed increased with an increase in lane width and in the presence of a median.

TOTAL TASK TIME, GLANCE DURATIONS AND NUMBER OF GLANCES

Total time spent by the drivers in performing a given task, number of glances made, and durations of individual glances provide information on how the drivers performed the task. Longer tasks times are indications of higher complexity in the tasks as well as effects of other driving and non-driving tasks that the driver shares with the tasks.

Table 15.1 provides data on total task times, mean glance durations and mean total number of glances from two reports (Rockwell, Bhise and Nemeth, 1973; Green and Shah, 2004). All the tasks included in the table were visual in nature. The speedometer reading and rearview mirror viewing tasks only involved the recording of driver

TABLE 15.1
Total Task Time, Glance Durations and Number of Glances in Performing In-Vehicle Visual Tasks

No.	Task Description	Mean Total Task Time (sec)	Mean Glance Duration (sec)	Mean Total Number of Glances	Reference
1	Speedometer Reading during Freeway Merging	NA	0.41-0.68	0.7-1.7	Rockwell, Bhise and Nemeth (1973)
2	Inside Rearview Mirror Viewing during Freeway Merging	NA	0.58-0.68	1.2-2.3	Rockwell, Bhise and Nemeth (1973)
3	Outside Rearview Mirror Viewing during Freeway Merging	NA	0.52-1.08	0.4-5.6	Rockwell, Bhise and Nemeth (1973)
4	Dialing a Phone Number (10 or 11 digits)	9.3-42.0	1.23-3.2	4.7-12.8	Green and Shah (2004)
5	Tuning a radio	7.0-27.5	0.67-2.87	2.0-15.0	Green and Shah (2004)
6	Entering Street Address	34.3-91.94	1.00-1.40	19.0-34.5	Green and Shah (2004)
7	Entering Destination (other than street address)	13.4-159.0	1.05-2.75	4.0-33.0	Green and Shah (2004)

eye and head movements. The radio and navigation system tasks involved hand and finger movements in operating the controls as well as eye movements needed to view the displays associated with devices. The total task time was defined as the time interval between the initiation of the driver's first response to perform the task until the end of the last response. The first response in a control activation task begins with the earlier of the two events to begin the task, namely, initiation of the hand movement from its prior position (from the steering wheel) or when the driver's eyes begin to turn to view the display. The last response will be the latter of the two events, namely, when the driver's hand (involved in operating the control) reached back to the steering wheel or when the eyes moved back to the road from the task related display. Different researchers have defined the total task time differently depending upon their ability to measure hand, finger, eye and head movements.

Tasks such as tuning a radio, entering a street address or destination into a device involve a number of steps with a sequence of control activations. These tasks, thus, involve a larger number of glances and longer total task completion times. The task completion times are also significantly affected by the driver age (Green and Shah, 2004). The older drivers may take twice the time required by the younger drivers. Green and Shah (2004) found that, while dialing a telephone number, the mean time to dial a digit (MTTDD) significantly increased with age and it could be predicted by the following equation: MTTDD = 0.55 + 0.039 (Age); where the time and the driver age were measured in seconds and years, respectively.

In another study, Jackson, Murphy and Bhise (2002) used the IVIS DEMAnD model (*In-Vehicle Information System Design Evaluation and Model of Attention Demand*) developed at the Virginia Polytechnic University's Transportation Research Center under the sponsorship of the Federal Highway Administration (Hankey, Dingus, Hanowski et al., 2001). The model was exercised to evaluate nine different tasks covering a range of attentional demands from a simple task such as glancing into a side view mirror to operating a complex navigation system. The nine tasks were simulated for three different vehicle configurations of interior instrument panel layouts (a center stack mounted LCD screen on top, middle and low locations) and

TABLE 15.2

Number of Glances and Total Task Times for Nine Different Tasks under Low and High Levels of Driver-Roadway Combinations

No.	In-Vehicle Task	Number of Glances		Total Task Time (s)	
		Low	High	Low	High
1	Check the driver's side mirror and locate an object present in the mirror field	1	2	1.1	2.5
2	Turn on the in-dash radio, select FM band, tune to specified frequency and adjust radio sound volume	3	7	7.2	18
3	Locate and remove a CD from the center console, remove a CD from its case, orient and insert in the player	3	8	6.5	21
4	Turn on a cellular phone, dial a 7-digit number and carry on a simple conversation with 9 questions	3	13	69	130.2
5	Search a specified travel route from a simple display of a in-dash navigation system	2	7	4.9	19
6	Identify required destination and current location, and select a desired route from the displayed list	3	10	10.3	25.5
7	Identify required destination and current location, evaluate route, and select a desired route	16	43	35.4	91.5
8	Listen to a variety of optional spoken routes and respond verbally when the desired route is mentioned	0	0	29.9	49.4
9	Verbally define desired location through a completely voice/speech technology interface	0	0	36.9	83.5

Source: Obtained from the IVIS DEMAnD Model.

two extreme levels of driver-roadway combinations (Low: Young driver [age 18–30] under low traffic density and low roadway complexity; High: Older driver [age 60+] under high traffic density and high roadway complexity). The predicted values of number of glances and total task times are summarized in Table 15.2.

DRIVER ERRORS

During driving tests, drivers can be asked to use a number of controls and displays and an experimenter seated in the front passenger seat, or a video camera located behind the driver can record errors made by the driver. The error data can be analyzed by creating measures such as error frequencies or error rates observed while driving under different conditions.

The errors (or difficulties encountered by the drivers) can be described as follows:

a) *Errors in locating, seeing and reading controls or displays*: for example, made two or more short glances or one long look, looked at wrong place, focused (or squinted) to see, and leaned to see.

b) *Errors in reaching and grasping*: for example, reached wrong place, leaned to operate, awkward grasp or orientation, missed the control, and accidental activation of a control.

c) *Errors in operating a control*: for example, moved in wrong direction, operated wrong control, exhibited trial and error during operation, repeated attempts made to use a control, overshot the control setting, looked again to verify a control setting, required an explanation from the experimenter to use the control or display and high effort in operating the control.

d) *Vehicle behavioral errors*: for example, changed speed (slowed down or increased speed by 5 mph or more), heading change, deviated in an adjacent lane, and abrupt heading correction.

SOME DRIVING PERFORMANCE MEASUREMENT APPLICATIONS

Table 15.3 presents several driving performance measurement applications and possible measures that can be used in evaluating the problems.

CONCLUDING REMARKS

Driver behavioral and performance measurements are used to evaluate various vehicle designs and features. The concepts introduced in this chapter allow ergonomics engineers to study problems related to vehicle evaluation and driver workload issues. Chapter 19 covers various evaluation methods used to determine effects of ergonomic changes and improvements in vehicle designs. The evaluations also allow for the comparison of different vehicle designs in terms of how they affect driver behavior and driver performance. Chapter 16 covers methods used in the measurement of driver workload, resulting driver distractions and their effects on driver behavior and performance.

TABLE 15.3
Description of the Driver Performance Measurement Applications and Possible Performance Measures

No.	Application Problem	Driver Performance Measures
1	Study on how drivers learn to drive	Standard deviation of lane position; mean velocity; standard deviation of velocity; eye fixations and glances made in different areas in the road scene, mirrors and in-vehicle devices.
2	Determine effectiveness of traffic control devices at freeway work zones	Standard deviation of lane position; mean velocity; standard deviation of velocity; eye fixations and glances made in different areas in the road scene.
3	Determine "good" car radio design	Total time spent in performing different radio tasks (e.g., turning in the radio, changing radio stations, changing bands and tuning a station, changing CDs); standard deviation of lane position; mean velocity; standard deviation of velocity; eye fixations and glances made on the radio.
4	Determine acceptable pedal layout	Percentage of driver ratings that the pedals are located "just right" using direction magnitude scales; percentage of drivers ratings that the pedal locations are acceptable.
5	Determine required field of view from a vehicle	Direction magnitude and acceptance ratings on a number of items, for example, pillar obscurations, pillar locations, down angle over the hood, up angle to view traffic signals, mirror locations, mirror width, mirror height, backlite, etc. (see Chapter 8 for more details).
6	Determine acceptable headlamp beam pattern	Seeing distances to pedestrian targets and lane lines; legibility distances to signs; discomfort glare ratings in opposed driving situations; ratings on perception of beam pattern on pavements (Jack et al., 1995).
7	Determine acceptable navigation system	Time to perform different tasks (e.g., destination entry, determining vehicle location); standard deviation of lane position; mean velocity; standard deviation of velocity; glances durations made in different areas in the road scene, mirrors and navigation screens; percentage of destinations correctly reached; NASA TLX Task Load Ratings.
8	Determine effective night vision system	Glance durations and percentage of time spent on the night vision screen and forward road scene; percentage of roadway targets detected; NASA TLX Workload Ratings.
9	Determine acceptable adaptive cruise control	Standard deviation of headway between lead vehicle and subject vehicle; standard deviation of relative velocity between lead vehicle and subject vehicle; NASA TLX Workload Ratings.

REFERENCES

American Association of State Highway Officials (AASHTO). 2005. *A Policy on Design Standards – Interstate System.* (called "The Green Book"). Fifth Edition, ISBN Number: 1-56051-291-1. Prepared by the Standing Committee on Highways, AASHTO Highway Subcommittee on Design, Technical Committee on Geometric Design, Washington, D.C., January 2005.

Fitzpatrick, K. et al. 2000. Design Factors That Affect Driver Speed on Suburban Arterials. Research Report 1769-3, Texas Transportation Institute, Lubbock, Texas.

Green, P. and R. Shah. 2004. Task Times and Glance Measures of the Use of Telematics: A Tabular Summary of the Literature. A Report on Safety Vehicles using adaptive Interface Technology (SAVE-IT, Task 6), the University of Michigan Transportation Research Institute, Ann Arbor, MI.

Green, P., B. Cullinane, B. Zylstra and D. Smith. 2004. Typical Values for Driving Performance with Emphasis on the Standard Deviation of Lane Position: A Summary of the Literature, A Report on Safety Vehicles using adaptive Interface Technology (SAVE-IT, Task 3A), the University of Michigan Transportation Research Institute, Ann Arbor, MI.

Hankey, J. M., T. Dingus, R. Hanowski, W. Wierwille and C. Anderws. 2001. In-Vehicle Information Systems Behavioral Model and Design Support: Final Report. Report No. FHWA-RD-00-135 sponsored by the Turner-Fairbank Highway Research Center of the Federal Highway Administration, Virginia Tech Transportation Institute, Blacksburg, VA.

Jack, D. D., S. M. O'Day and V. D. Bhise.1995. *Headlight Beam Pattern Evaluation – Customer to Engineer to Customer – A Continuation.* SAE paper no. 950592. Presented at the 1995 SAE International Congress, Detroit, MI.

Jackson, D., J. Murphy and V. D. Bhise. 2002. *An Evaluation of the IVIS-DEMAnD Driver Attention Demand Model.* Paper presented at the 2002 Annual Congress of the Society of Automotive Engineers, Inc., Detroit, MI.

Lambert, S., S. Rollins and V. Bhise. 2005. *Effects of Common Driver Induced Distraction Task on Driver Performance and Glance Behavior.* A paper presented at the 2005 Annual Meeting of the Transportation Research Board, Washington, D.C.

Rockwell, T. H., V. D. Bhise, and Z. A. Nemeth. 1973. Development of a Computer Based Tool for Evaluating Visual Field Requirements of Vehicles in Merging and Intersection Situations. Vehicle Research Institute Report, Published by the Society of Automotive Engineers, Inc., New York, NY.

Savino, M. R. 2009. Standardized names and Definitions for Driving Performance Measures. M.S. Thesis, Tufts University.

Tarrants, W. E. 1980. *The Measurement of Safety Performance.* New York, NY: Garland STPM Press.

Tijerina, L., E. Parmer, and M. J. Goodman. 1998. Driver Workload Assessment of Route Guidance System Destination Entry While Driving: A Test Track Study. *Proceedings of the 5th ITS World Congress*, Soul, Korea.

USDOT, 2003. *Manual of Uniform Traffic Control Devices for Streets and Highways.* Prepared by the Federal Highway Administration.

16 Driver Workload Measurements

INTRODUCTION TO DRIVER WORKLOAD

The driver workload measurement is a specialized topic in vehicle ergonomics that is used in developing and applying methods to measure the total amount of mental workload due to the tasks that a driver performs under any given situation. This topic is of particular importance as technological advances are generating new features to enable the drivers to access and use more information from various in-vehicle devices while driving. To answer the question of how many different tasks can a driver safely perform simultaneously in a given situation, thus, assumes that we are able to measure the driver's total workload and compare it with the driver's capabilities, and determine if the driver has sufficient spare capacity left for any emergencies or is overloaded.

The problem becomes more complex because both the driver's total workload and driver's capabilities (or capacities) to perform the driving and non-driving tasks are not constant during driving. The total number of tasks that the driver needs to perform at a given instant include (a) involuntary tasks, which cannot be performed under the driver's discretion as these tasks are demanded by changing situations (e.g., variations in traffic, roadway, weather) that are outside the driver's control and must be performed under certain time pressure, and (b) other voluntary tasks, that can be performed at the driver's discretion.

The objectives of this chapter are to review various approaches and methods available to measure and evaluate the driver's workload.

DRIVER TASKS AND WORKLOAD ASSESSMENT

The driver's total workload includes all driving and non-driving tasks:

1. *Normal driving tasks (involuntary)*: monitoring the roadway and making lateral and longitudinal control actions by using primary vehicle controls (i.e., steering wheel and pedals) and other safety related driving controls (e.g., defrosting windshield) and displays

DOI: 10.1201/9781003485605-3

2. *Situational tasks (involuntary)*: responding to the demands from the roadway (e.g., curves, lane drops, merges) and the traffic (e.g., changing lanes, passing) and other demands (e.g., avoiding colliding with a crossing pedestrian)

3. *In-vehicle tasks (mostly voluntary)*: reading displays and operating controls and use of secondary interfaces (e.g., climate controls, entertainment devices)

4. *Other voluntary tasks*: talking with other passengers, reading maps, using cell phones, reading notes, drinking beverages, eating, attending to other passenger needs, and so forth.

The above tasks require mental and physical work. The mental work involves information acquisition (involving sensing, detecting, recognizing) and processing information (searching, discriminating, selecting and integrating sensed information from different modalities, analyzing, retrieval/storage of information in human memory systems, and decision making) and executing and making control actions. Physical work includes generation of muscular forces to produce coordinated movements of different body parts (e.g., head, hand, arm, foot, leg and torso) with needed speeds and accuracy.

The workload assessment, thus, involves measuring the driver's workload and answering questions such as: How busy is the driver? How many tasks can the driver handle safely? Would the driver be overloaded? The key considerations here are (1) the driver has limited capabilities to perform tasks, and (2) if the demands of the tasks are greater than the driver's capabilities, then (a) the driver may perform the tasks but experience higher stress; (b) the driver may make errors; (c) the driver may slow down; (d) the driver may not perform some tasks or parts of some tasks; or (e) some combination of above depending upon priorities and capabilities of the driver.

PRESENT SITUATION IN THE INDUSTRY

Despite considerable amount of research reported in the literature in this area and because of the complexities associated with the combination of different mental and physical tasks associated while driving, there is currently no single accepted or recognized method in the automotive industry to decide on what new in-vehicle devices and features are safe enough to be incorporated in a production vehicle. Thus, automobile manufacturers resort to applications of multiple approaches and methods to evaluate the new devices. The positive and negative aspects of the information gathered in these evaluations are then reviewed and discussed with different levels of subject matter experts, management and legal experts to determine if a product feature or a device should be incorporated in a production vehicle.

CONCEPTS UNDERLYING MENTAL WORKLOAD

The concept of mental workload has been studied and is being constantly researched in the field of cognitive psychology. Many researchers have used the concept to explain how the human operator processes information to perform complex tasks and developed models to explain human performance and to measure workload (Meshkati, Hancock and Rahimi, 1992; Tijerina, Parmer, and Goodman, 2000; SAE, 2009; ISO, 2008). Realizing that the area of mental models and workload measurements are evolving, the following brief statements will summarize approaches and concepts used in understanding the concept of metal workload.

1. Humans have limited information processing capacity. Simple models of information processing have shown that reaction times can be used to measure the driver's information processing capacity (see Volume 1, Chapter 6). The amount of processing capacity of the driver does not remain constant during driving due to changes in a number of factors such as attention, distraction, fatigue, roadway, and traffic situations.

2. Humans have multiple resources to process information from different modalities such as vision and audition. In addition, a central resource is shared and used for processes such as spatial organization of acquired data, verbal processing, and response selection. The central processor can act as a serial, parallel or hybrid processor depending upon the operator characteristics such as practice, experience, stress, attention level, and age (Wickens, 1992).

3. The mental workload is the specification of the amount of information processing capacity that is used for task performance. Demand is determined by the goal that must be attained by means of task performance.

4. Workload can be defined as that portion of the driver's limited capacity that is required to perform a particular task.

5. Workload is not only task-specific but is also person-specific. Thus, each individual would have a different workload while performing the same set of tasks.

6. Workload is not an inherent property, but rather it emerges from the interaction between the requirements of a task, the circumstances under which it is performed, and the skills, behaviors, and the perception of the operator (Hart and Staveland, 1988).

7. The complexity associated with a task will increase with an increase in the number of stages of processing that are required to perform the task.

8. The difficulty of a task is related to the processing effort (e.g., number of resources consciously allocated) that is required by the individual for task performance.

9. Spare capacity can be defined as the additional or excess information processing capacity available at any given instant. Under most driving situations drivers have considerable spare capacity available. However, addition of tasks can reduce or even eliminate the spare capacity and leave the driver with insufficient capacity to perform the required amount of information processing.

10. The term visual spare capacity is used to refer only to the unused capacity of the visual resource available to process visual information acquired through the driver's eyes. An occlusion task which involves controlling the durations "open" (i.e., vision is available) and "closed" (vision occluded) is one of the methods used to measure visual spare capacity (see Volume 1, Chapter 6).

11. Many factors affect the driver's workload. The factors can be categorized as: (a) driver state affecting factors: e.g., monotony, fatigue, sedative drugs and alcohol; (b) driver trait factors: for example, experience, skills, age, and strategy; (c) environmental factors: for example, road environmental demands, traffic demands and weather conditions; and (d) vehicle factors: for example, driver–vehicle interface, automation, and feedback provided by the vehicle. Thus, performance, effort and spare capacity may or may not be related because of variations in the driver's skills, motivation, attention, and capabilities.

12. Most driver failures occur due to information processing failures, that is, the inability of the driver to make the right decision at the right time and right place.

METHODS TO MEASURE DRIVER WORKLOAD

1. *Driver Performance Measurements*: Various driver performance measurement methods and measures can be used to determine the effect of the driver workload on driver performance in performing different tasks (see Chapter 15). The changes in levels of performance measures obtained while the driver is simply driving (i.e., primary baseline driving tasks) versus when the driver is asked to perform tasks in addition to driving (i.e., secondary or dual task, e.g., driving and dialing a cell phone) have been used to assess the effects of driver workload. Table 16.1 presents results from four studies showing worsening of driver performance by a factor of 1.12 to 1.88 when the drivers were asked to perform various distracting secondary tasks in addition to primary driving task.

2. *Physiological Measurements*: These measures assume that the workload will affect bodily functions. They are based on measuring effects of arousal, excitement, stress/tenseness, thought processes, and use of body movements through muscle activations that are caused during performance of the tasks.

A number of physiological measurements have been used to determine the effects of the workload on the human operator. Some examples of the physiological measurements are heart rate, respiration rate, brain's spontaneous electrical activity from electroencephalograms (EEG) and evoked potential recordings, electrical activity of the heart from electrocardiograms (EKG), electrical activity of muscles from electromyograms (EMG), electrical activities of the eye muscles from electro-oscillograms (EOG), galvanic skin response (GSR), body and skin temperatures, sweat rate, pupil size and eye blink rate.

The variations in the body functions due to extensive physical biomechanical workload (mostly in industrial tasks) can be measured using heart rate, respiration rate, oxygen intake and EMG. These measures related to muscular activities are easier to measure and interpret as compared to variations due to changes in the mental workload during driving. Brookhuis and Waard (2000) have reported studies that showed

TABLE 16.1
Illustrative Results from Four Performance-Based Studies

Sr. No.	Task Description		Measure	Baseline: Driving only Single Task (S)	Driving and Secondary Task Dual Task (D)	Ratio of Dual Task to Single Task (%) (D/S)%	Reference
	Primary Task	Secondary Task					
1	Car following on a freeway	Exchanging text messages	Std. Dev. Of Following Distance	11.9 m	17.9 m	1.5	Drews, Yazdani, Godfrey, Cooper & Strayer, 2009
			Lane crossings per km	0.26	0.49	1.88	
2	Driving on a simulated 2-lane roadway at about 50 mph	Reading cross streets on a map	Std. Dev. of Lateral Position	0.27 m	0.41 m	1.5	Lambert, Rollins & Bhise, 2005
		Read a short message from a text pager	Std. Dev. of Lateral Position	0.34 m	0.47 m	1.41	
3	Driving on a simulated roadway at 60 kph	Sending text message using cell phone	Mean Time headway	4.3 s	6.4 s	1.49	Hosking, Young & Regan, 2009
			Std. Dev. of Lateral Position	0.2 m	0.29 m	1.45	
4	Following a lead vehicle while braking from 65 mph to 30–45 mph in a fixed-base driving simulator	Naturalistic conversing on a hands-free cell phone with a research confederate after 4 days of practice	Collision frequency (vehicle contacted objects in the environment)	7	12	1.71	Cooper and Strayer, 2008
			Brake reaction time	1.00 s	1.20 s	1.2	
			Following distance	22.5 m	25.1 m	1.12	

Source: All figures and tables are created by the author.

that the drivers' heart rate increased under higher stress and workload, for example, driving through traffic circles increased heart rate as compared to straight roads, and awaiting traffic light increased heart rate variability. Mehler, Reimer, Coughlin and Dusek (2009) examined the sensitivity of heart rate, skin conductance, and respiration rate as measures of mental workload in simulated driving environment using a sample of 121 young adults. Their results showed that as the mental workload imposed by "n-back task" (recalling a one digit number from 0, 1 or 2 back positions) increased the heart rate, the skin conductance and the respiration rate. Verwey and Zaidel (2000) found that the eye blink rate was related to drowsiness, suggesting that the frequency of eye closures exceeding 1 s are indicators of drowsiness.

The reliability of physiological measures in measuring the driver's mental workload is poor because of large variations among individuals and many factors (e.g., anxiety, stress) that affect these measures. The physiological measures are rarely used during the vehicle development process because (a) the links between the physiological measures and performance are not clearly understood, and (b) the high expense and time associated with collecting large amounts of data and analyses requiring more sophisticated techniques.

3. *Subjective Assessments*: Subjective ratings on the level of difficulty, stress, discomfort, mental workload, and physical workload provided by subjects during performance of different tasks are commonly used as measures to assess the workload. Three well-developed subjective workload measurement techniques used in this field are (1) the NASA Task Load Index (TLX); (2) the Subjective Workload Assessment Technique (SWAT); and (3) the Workload Profile (WP).

The NASA-TLX is a multi-dimensional rating procedure that derives an overall workload score based on a weighted average of ratings on six subscales. These subscales include Mental Demand, Physical Demand, Temporal Demand, Own Performance, Effort and Frustration. The method has been used to assess workload in various human-machine environments, such as aircraft cockpits, workstations, process control environments and various actual driving as well as in simulated driving situations (Bhise and Bhardwaj, 2008).

The subscales can be specified by 5-point, 10-point or 100-point interval scales. The standardized descriptors (questions) used for each subscale category and adjectives used to define their end points are presented in Table 16.2. The ratings on individual subscales can be used as evaluation scores or an overall workload score can be obtained from a weighted sum of the ratings on the six scales. The weightings of the subscales can be obtained after the subjects have performed all the tasks. A paired comparison method (see Chapter 19) can be used to obtain the weightings based on the importance of the subscale categories associated with the tasks.

The SWAT involves asking the operator to rate workload using three scales, namely, Time Load, Mental Effort Load and Psychological Stress Load. Each scale has three levels: Low, Medium and High. The descriptors used to define the three levels of each of the three scales are presented in Table 16.3. The method uses conjoint

TABLE 16.2
Description of the Six Subscales Used in the Measurements of the NASA Task Load Index (TLX)

No.	Subscale (Workload Attribute)	Adjectives used to describe the low and high end points of the scale	Questions used to describe the scale attribute
1	Mental Demand	Low, High	How much mental and perceptual activity was required (e.g., thinking, deciding, calculating, remembering, looking, searching, etc.)? Was the task easy or demanding, simple or complex, exacting or forgiving?
2	Physical Demand	Low, High	How much physical activity was required (e.g., pushing, pulling, turning, controlling, activating, etc.)? Was the task easy or demanding, slow or brisk, slack or strenuous, restful or laborious?
3	Temporal Demand	Low, High	How much time pressure did you feel due to the rate or pace at which the tasks of task elements occurred? Was the pace slow and leisurely or rapid and frantic?
4	Performance	Good, Poor	How successful do you think you were in accomplishing the goals of the task set by the experimenter (or yourself)? How satisfied were you with your performance in accomplishing these goals?
5	Effort	Low, High	How hard did you have to work (mentally and physically) to accomplish your level of performance?
6	Frustration Level	Low, High	How insecure, discouraged, irritated, stressed and annoyed versus secure, gratified, content, relaxed and complacent did you feel during the task?

measurements and scaling techniques to develop a single interval rating scale (Reid and Nygren, 1988).

The workload profile (WP) method is based on the multiple resource model (Wickens, 1998). It considers the following eight workload dimensions as attentional resources (1) perceptual/central processing; (2) response selection and execution; (3) spatial processing; (4) verbal processing; (5) visual processing; (6) auditory processing; (7) manual output; and (8) speech output. The subjects are asked to provide proportions for each of the eight workload dimensions used in each task (in a random order) after they have experienced all the tasks. Thus, each task is evaluated by providing eight ratings, each between 0 to 1 to represent the proportion of each attentional resource used in the task. Thus, a rating of "0" means that the task did not

TABLE 16.3
Descriptors Used to Define Three Levels of the Time Load, Mental Effort Load and Psychological Stress Load Scales Used in the Subjective Workload Assessment Technique (SWAT)

No.	Scale	Level	Descriptors
1	Time Load	Low	Often have spare time. Interruptions or overlap among activities occur infrequently or not at all.
		Medium	Occasionally have spare time. Interruptions or overlap among activities occur infrequently.
		High	Almost never have spare time. Interruptions or overlap among activities are very frequent, or occur all the time.
2	Mental Effort Load	Low	Very little conscious mental effort or concentration required. Activity is almost automatic, requiring little or no attention.
		Medium	Moderate conscious mental effort or concentration required. Complexity of activity is moderately high due to uncertainty, unpredictability, or unfamiliarity. Considerable attention required.
		High	Extensive mental effort and concentration are necessary. Very complex activity requiring total attention.
3	Psychological Stress Load	Low	Little confusion, risk, frustration, or anxiety exists and can be easily accommodated.
		Medium	Moderate stress due to confusion, frustration, or anxiety noticeably adds to workload. Significant compensation is required to maintain adequate performance.
		High	High to very intense stress due to confusion, frustration, or anxiety. High extreme determination and self-control required.

require the dimension and "1" means that the task required maximum attention. The ratings on the eight dimensions of each task are later summed to obtain an overall workload rating for the task.

The subjective methods are commonly used in vehicle development because they are easier to obtain, require no instrumentation and have high "face validity". The disadvantages of the subjective methods are that the rater may find it difficult to understand many issues associated with comparing different products and situations; and the agreement between different raters may not be unanimous. Chapter 19 provides additional information on subjective methods used in the industry.

4. *Secondary Task Performance Measurement*: This approach uses an artificially added task called a "secondary task" while performing a primary task and assumes that an upper limit exists on the capacity of the driver to gather

and process information. The performance in the secondary task is used to measure the driver's workload on the assumption that adding the secondary task on to top of the primary driving task (e.g., lateral and longitudinal control of the vehicle) will increase the driver's total workload; and if the secondary task is sufficiently difficult, the driver would reach or exceed his overall capacity to perform both the tasks. By carefully controlling the driver's priorities, by emphasizing during instructions (e.g., the driver must maintain constant 100 km/h speed and keep the vehicle in the lane all the time); the driver will be asked to maintain performance in the primary task. Thus, the level of performance in the secondary task will indicate the amount of workload or capacity taken up by the primary task.

Many different secondary tasks have been used in literature. Some examples of secondary tasks are peripheral detection tasks, arithmetic addition tasks, repetitive tapping tasks, time estimation, random number generation, choice reaction time tasks, critical tracking tasks, visual search tasks, memory search tasks, and so forth. For example, Olsson and Burns (2000) measured reaction time to detect peripheral targets and hit rate (proportion of targets correctly detected) while driving a car under a baseline driving condition and in other conditions involving the baseline driving and performing radio station tuning and CD tasks (e.g., turn on the CD mode and select track). They found that when the drivers were asked to perform these additional radio and CD tasks, their peripheral detection performance, measured by both the reaction time and hit rate, were degraded as compared to their performance in the baseline driving condition.

The use of the secondary task as a method to measure driver workload has a number of shortcomings. If the introduction of the secondary task modifies or interferes with the driver's primary task, then the driver may be forced to change his strategy, and thus, may distort the load imposed by the primary task. The interference in the primary task is greater when the tasks share the same response resources than if the responses occupy different resources. Further, it is difficult for some drivers to maintain the same level of attention and priority in performing the primary task when the secondary task is introduced.

5. *ISO Lane Change Test* (LCT): The lane change task proposed in the ISO draft document (ISO, 2008) is a simple laboratory dynamic dual task method that quantitatively measures performance degradations in a primary driving task while a secondary task is being performed. The primary task involves driving on a 3 km length straight 3-lane roadway at 60 km/h constant speed and making a series of lane changes as indicated on pairs of signs located on the road shoulders. The mean distance between the pairs of signs is 150 m. Thus, the driver is forced to make 20 lane changes during the 3 km drive. The driver sits in front of a steering wheel mounted on a table and views the driving scene in a video monitor located on the table (see Figure 16.1). A number of secondary tasks can be used for the driver to perform while performing the above-described primary lane change task. The entire procedure, including

its software, test track, sign configuration, laboratory set-up, data collection, primary and secondary task performance measures and analysis procedure are standardized and specified in the ISO document.

The primary task performance measure is the average deviation in the path of the vehicle, called the MDEV (mean deviation). It is calculated with respect to a reference path trajectory. Two methods are provided for selecting the reference trajectory. The first method called the "adaptive model" method uses the same subject's trajectory obtained at the end of the practice session under the baseline condition (which involves performing the primary driving task with the lane changes only, i.e., without a secondary task). The second method, called the "normative model" method, uses the reference trajectory as a "nominal trajectory" specified under an idealistic set of assumptions (to simulate an ideal driver who changes lanes with 0.6 s reaction time and drives exactly at the center of the driving lane) and it is the same of all subjects. The use of the normative model method is optional.

The LCT procedure has been used by a number of researchers to evaluate its repeatability, consistency and usefulness. The method was used by Jothi (2009) and the study is described later in this chapter.

6. *IVIS DEMAnD Model*: A computer model known as the "In-Vehicle Information System Design Evaluation and Model of Attention Demand" (called the *IVIS DEMAnD Model*) can be used to predict the level of driver distraction associated with an in-vehicle device. The model was developed by Hankey, Dingus, Hanowski, Wierwille and Anderws (2001) to allow automotive designers and safety engineers to predict driver workload levels on either an existing or a proposed vehicle information system. Using this model, an engineer can enter dimensional data for a given vehicle, along with the locations of controls and displays for the system being analyzed, and the model uses empirical data from a database of human factors to predict the driver workload associated with the system or a task completed on that system.

Jackson and Bhise (2002) exercised the model under nine different driver-attention task levels ranging from a simple task, such as glancing into a side view mirror, to a very complex task, such as operating an in-vehicle navigation system. The nine driver tasks were evaluated using three different vehicle configurations and two levels of driver-roadway complexity. In addition, a real-world study was conducted to measure the drivers' visual performance using four of the above nine tasks for comparison with model predictions. The comparisons of model predictions of maximum number of glances, total task performance time, and a model-rating feature called the *figure of demand* for each of the tasks showed the following: (1) Driver task performance behavior was influenced substantially more by the differences in the level of driver/roadway/traffic combinations, than by the differences in the test vehicle configurations. (2) The driver performance data collected under the actual driving conditions compared very well with the model predictions. (3) Overall, the

IVIS DEMAnD model was found to be a good early attempt at modeling the effect on the driver performance using the in-vehicle information systems.

7. *SAE J2364 and J2365 Recommended Practices*: The Society of Automotive Engineers published two recommended practices, namely *SAE J2364* and *SAE J2365*, to measure driver workload for the navigation and route guidance system tasks (SAE, 2009a,b).

The SAE J2364 practice presents a laboratory-based procedure to measure total time taken by subjects to perform a given task related to a navigation and route guidance system. The method, thus, assumes that a working model of the in-vehicle navigation and route guidance system is available for the laboratory test. Two methods of obtaining total time can be used. The first method requires the driver to do the entire task uninterrupted, whereas the second method allows the use of an occlusion device. The subjects for the test should be licensed older drivers (age 45 to 65) who are initially not familiar with the device. The total time taken to complete the task is measured after 5 practice trials. The SAE practice recommends using 10 subjects and measuring total time three times (3 trials after the first 5 practice trials) for each subject. The task is performed under a static non-driving situation in a laboratory bench test type situation. For the occlusion method, the total time is obtained by summing all the glances made on the device to perform the task. The task is considered to be acceptable if it can be completed in less than 15 s. This method is known in the industry as the SAE's "15-second Rule".

The SAE J2365 practice presents an analytical method to calculate the time to complete a given in-vehicle navigation and route guidance task. The advantage of this method is that it does not require a working version of the device and it can be used in an early design phase. The calculation method is based on the goals, operators, methods and selection rules (GOMS) model described by Card, Morgan and Newell (1980, 1983). The method essentially involves: (a) breaking down the task into a series of simple steps; (b) applying predetermined times by considering appropriate mental, key strokes, age multipliers and other operators in each step; (c) summing the assigned times for each sub-goal (e.g., move hand to the device, select a city) and goal (e.g., enter a destination using the street address method) in each step; and (d) summing the completion times over all steps to obtain an estimate of the total time required. This method is especially useful in the early design phase, where a number of alternate design concepts and their operational features can be evaluated by comparing total time estimates to perform each given task. The acceptability of the design can be also judged by comparing the total time with a pre-selected requirement, such as the 15-second rule. This method does not consider voice-activated controls, voice outputs and communication between the driver and others, or passenger operation.

SOME STUDIES ILLUSTRATING DRIVER WORKLOAD MEASUREMENTS

The most common approach used by many researchers involves asking subjects to drive under a baseline driving situation (when the subjects are only driving and not performing any other secondary tasks) and comparing their performance in the baseline situation with the performance obtained while performing dual tasks (i.e., performing the baseline task and additional secondary task). Various secondary tasks have been used by different researchers while performing driving tasks under different baseline situations (e.g., open road driving, car following, lane changing). Five different studies reported in the literature are presented below.

DESTINATION ENTRY IN NAVIGATION SYSTEMS

Tijerina, Parmer and M. J. Goodman (1999) evaluated four commercially available navigation systems in a test car. Sixteen test participants (half males and half females; half below age 35 and half above 55) drove a 1993 Toyota Camry with microDAS (Data Acquisition System) that captured travel speed, lane position, and lane exceedances, as well as video of road scene and eye glance behavior at a sampling rate of 30 Hz. The participants drove the vehicle on a 12 km (7.5 miles) multi-lane oval track at speeds between 72–96 km/h (45–60 mph). The participants were asked to enter point of interest destinations in each of the four systems (tested in random order) along with other two tasks which included (a) dialing an unfamiliar 10-digit phone number in a handheld cell phone, and (b) entering a radio station manually in a modern radio.

The response measures used in the study included (a) total time to complete the task; (b) mean glance frequency to the device; (c) average glance duration; (d) average total eyes-off-the-road time; and (e) number of lane exceedances during each task trial.

The results showed the following:

(a) The average total time spent in destination entry was 68 s for the younger drivers and 118 s for the older drivers. Whereas the cell phone 10-digit entry and the radio tuning tasks took on average about 25 and 22 s, respectively.

(b) The mean number of glances made by the subjects in the visual-manual destination entry tasks ranged from 22 to 33. Whereas the glances made during audio entry of the destinations, cell phone 10-digit dialing, and radio tuning took 4, 8 and 6 glances, respectively.

(c) The average glance duration during visual-manual entry tasks in all the devices ranges between about 2.5 to 3.2 s.

(d) The total eyes-off-the-road time ranged between 60 to 90 s for the visual-manual destination entry tasks; and the cell phone 10-digit dialing, and radio tuning tasks took, on average, 17 and 16 s, respectively.

(e) The average number of lane exceedances per trial during the destination entry tasks ranged from 0.2 to 0.9; whereas the cell phone dialing and radio tuning tasks had 0.06 and 0.2 lane exceedances per trial, respectively.

HANDHELD VERSUS VOICE INTERFACES FOR CELL PHONES AND MP3 PLAYERS

Owens, McLaughlin and J. Sudweeks (2010) conducted an on-road study to compare driver performance in using handheld versus in-car voice control interfaces. They asked 21 participants to drive a test vehicle on straight sections of a divided secondary road with 105 km/h (65 mph) speed limit and asked the drivers to perform the following tasks: (1) baseline: only driving at the beginning and end of the study; (2) dial a contact person's number; (3) converse on the phone with an experimenter on a predetermined topic of interest; (4) play a music track. The tasks 2 to 4 were performed by using two types of equipment: (1) a personal handheld cell phone and a personal MP3 player; and (2) the SYNC voice control interface in a 2010 Mercury Mariner (mid-size SUV).

They used an onboard data acquisition system to measure steering wheel position, and video cameras were used to measure number of glances on the user interface, maximum glance duration and task duration. After each task, the participants were also asked to provide a rating on mental demand using a 1 to 7 scale (1=low demand, 7= high demand).

The results of the study showed this:

(a) The task duration times were significantly shorter while using the voice interface as compared to the handheld devices. For dialing a 10-digit phone number, the voice interface took about 10 s as compared to about 15–20 s using the handheld cell phone. The track play tasks took about 10 s with the voice interface as compared to about 35–40 s using the handheld MP3 players.

(b) The steering variance in the baseline condition was about 1.0 deg^2. The variance increased while using both the phone and MP3 handheld devices from about 1.2 to 1.5 deg^2 for the younger drivers to 2.0 deg^2 for the older drivers. The steering variance during the use of voice interface did not increase from the baseline.

(c) The number of glances made during the voice interface usages were on average about 8–9 for the dialing and 15–20 for the track playing tasks using the handheld devices. With the voice interface, the average number of glances for the two types of tasks ranged between 3–4 glances.

(d) The mental demand ratings for the handheld devices ranged from about 3 to 5.5 as compared to 2 to 3 for the voice interface. The mental demand ratings for the older drivers were always higher by 1.5 to 2 points as compared to the younger drivers. The mental demand ratings for the baseline condition were on average 1.5 and 2.0 points for the younger and older drivers, respectively.

TEXT MESSAGING DURING SIMULATED DRIVING

Drews, Yazdani, Godfrey, Cooper and Strayer (2009) evaluated the effects of text messaging while driving a fixed-base driving simulator with three screens, providing approximately 180 forward degree field of view. They compared the performance of forty participants while driving (single task) and driving and text messaging (dual task). The participants drove in the right lane of a 51 km (32-mile) simulated multilane rural and urban beltway. During each trial, the participants followed a pace car that was programmed to brake at 42 randomly selected intervals and would continue to decelerate until the participant depressed the brake pedal, at which point the pace car would begin to accelerate to the normal freeway speed. During the dual task condition, the participants used their own cellular phones and received and composed text messages.

The driving performance was measured by using (a) brake onset time (reaction time of the following driver); (b) following distance; (c) standard deviation of the following distance; (d) minimal following distance; (e) lane crossings per kilometer; (f) lane reversals per kilometer; and (g) gross lateral displacement.

Their results showed the following:

(a) Mean brake onset time increased from 0.88 s in the baseline driving to 1.08 s in the dual task driving.
(b) The average following distance increased from 29.1 m in the baseline driving to 34.3 m in the dual task driving.
(c) The minimal following distance decreased from 9.0 m in the baseline driving to 6.8 m in the dual task driving.
(d) The lane crossing frequency, lane reversal frequency and the gross lateral displacement increased in the dual task driving as compared to the baseline driving by 88 percent, 26 percent and 26 percent, respectively. The authors, thus, concluded that text messaging during driving has a negative impact on simulated driving.

COMPARISON OF DRIVER BEHAVIOR AND PERFORMANCE IN TWO DRIVING SIMULATORS

Bhise and Bhardwaj (2008) conducted a study to compare driving behavior and performance of drivers in two different fixed-base driving simulators (namely, FAAC and STI) while performing the same set of distracting tasks under geometrically similar freeway and traffic conditions. The FAAC simulator had a wider three-screen road view with steering feedback as compared to the STI simulator which had a single screen and a narrower road view and had no steering feedback. Twenty-four subjects (12 younger and 12 mature) drove each simulator on a freeway type roadway with geometrically similar characteristics and were asked to perform a set of nine different tasks involving different distracting elements.

The nine tasks were as follows:

1) Collect Ontario map from the map compartment (on the lower part of the driver's door).
2) Answer the cell phone (after the subject's own cell phone ,placed in the center console, rang).
3) Collect 65 cents from the coin holder located in the center console.
4) Switch (radio) to FM and tune to preset # 4.
5) Collect two yellow and two red candies (placed in the console).
6) Sip the beverage (water bottle) from the cup holder (located in the front part of the center console).
7) Search for the keys located in the center console (inside compartment with a hinged lid).
8) Check if there is voice mail on the cell phone.
9) Replace the CD in the CD player (remove the CD and insert a different CD from a CD case placed on the center console).

The following performance measures were obtained:

1) Number of glances made away from the forward scene to complete the task
2) Longest glance duration
3) Total task completion time
4) Number of lane deviations during task completion
5) Maximum lane deviation
6) Vehicle speed
7) Number of crashes
8) Average mental demand rating
9) Average physical demand rating
10) Average temporal demand rating
11) Average performance demand rating
12) Average effort rating
13) Average frustration rating

The results showed the following:

a) Driver behavioral measures, such as number of glances made in performing a task, total task completing time, and the NASA TLX workload ratings (on 1–10 scales) differed due to the differences in the tasks. However, the behavioral measures and the NASA TLX ratings showed remarkably similar behavior in the two simulators.
b) Overall, Tasks #4 (Switch (radio) to FM and tune to preset # 4) and task #7 (Search for the keys located in the center console) were the least demanding; whereas, Task #3 (Collect 65 cents from the coin holder located in the center console) and Task#8 (Check if there is voice mail on the cell phone) were the most demanding.
c) The drivers' driving performance, measured by maximum lane deviation, average speed, and number of accidents, were significantly different in the two simulators.
d) The driver performance was significantly better while driving the FAAC simulator than the STI simulator.
e) The results, thus, showed that while the demand placed on the drivers due to the distracting tasks produced similar glance behavior and task loadings in the two simulators, the narrower road view and lack of steering feedback in the STI simulator produced substantially degraded driving performance than the performance observed in the FAAC simulator.

APPLICATIONS OF THE ISO LANE CHANGE TEST

The lane change test (LCT) proposed by the ISO/TC2/SC22 (ISO, 2008) was conducted in 2008 at different sites in different countries to test the calibration and replication capability of the method. One of the test sites was the Vehicle Ergonomics Laboratory at the University of Michigan–Dearborn campus. The study was conducted by Jothi

(2009) under the guidance of the Alliance of Automotive Manufacturers (AAM) members and Profs. Bhise and Rodrick at the university.

The test set-up is presented in Figure 16.1. The primary task of the driver during the entire testing was to drive the lane change course at a constant 60 km/h and make quick lane changes as indicated by each pair of signs. The driver's view of the lane change course presented on the screen is shown in Figure 16.2.

Twenty-four subjects participated in the study. After extensive familiarization with the driving simulator and the lane-change procedure, each subject was asked

FIGURE 16.1 The laboratory set-up of the LCT test showing the driver's screen on the left and the secondary task screen and controls on the right side table.

FIGURE 16.2 The driver's view of the LCT screen showing a series of side mounted signs indicating driving lane.

to drive the 3 km route ten times and perform the lane change task. The first and last (tenth) runs involved driving only with the lane changes (single task) and thus they were called Baseline 1 and Baseline 2, respectively. The middle eight runs were randomly assigned to the dual tasks involving four types of secondary tasks. Each secondary task had two levels – Easy and Hard. And the two levels of tasks for each type were conducted sequentially in separate runs. The secondary tasks were as follows: (1) Critical Tracking Task (Easy and Hard); (2) Visual Search Task (Easy and Hard levels); (3) Sternberg Memory Task (Easy and Hard levels); and (4) Nomadic Task with a TomTom Navigation System (Easy and Hard levels). During the dual task runs, the subjects were asked to perform the selected secondary task continuously, in repeated trials, throughout the entire run while performing the primary task of driving and constantly changing the lanes.

The critical tracking task (CTT) involved stabilizing an increasingly unstable vertically moving bar within a marked interval shown in the right-hand screen. The subject controlled the bar by using two up and down arrow keys of a keypad placed on the right hand side of the table. The stability parameters were changed to create easy and hard tracking difficulty levels.

The visual search task (called SURT – Surrogate reference task) presented in the right-hand screen involved detecting a larger target ring among many scattered smaller background rings with identical size. The number of background rings and the difference between the diameter of the target and the background rings were varied to create easy and difficult tasks. The easy task had a larger diameter target ring and a smaller number of background rings as compared to the difficult level.

The Sternberg task (called COTA – Cognitive task) involved the subject to first listen to a set of 3 (for easy level) or 6 (for hard level) randomly selected single digit numbers and then was presented another probe digit. The subject had to determine if the probe digit was included in the set of digits presented earlier and respond by pushing "Yes" or "No" keys placed on the right side table.

The nomadic task involved the subject selecting a required screen of a TomTom navigational system placed on the right side table and adjusting the sound volume to the required level provided by the experimenter for the easy level. For the hard level, the subject was given a destination address on a 76 x 127 mm (3"x5") card and was asked to enter the city name, street name and the street number.

The data collection software was programmed to store data for the primary and secondary tasks for all the runs. The stored data were analyzed to obtain mean difference in lateral deviation trajectory (MDEV) using both the adaptive and the normative models for each run of each subject. The adaptive model used the subject's qualifying lateral position profile obtained at the end of the practice runs. The normative model used the basic lateral profile obtained by assuming that an ideal driver makes all the lane changes with 0.6 s response time. The mean values of the lateral deviations obtained from the data over all the 24 subjects for the 10 runs are shown in Figure 16.3.

From Figure 16.3, the following observations can be made:

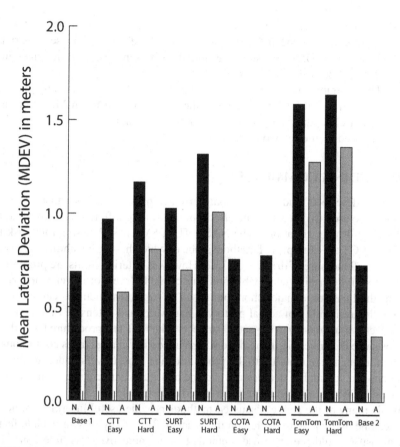

FIGURE 16.3 Mean lateral deviations (MDEV) values obtained by using the normative (N) and the adaptive (A) models for the baseline (called Base 1 and Base 2) and dual task conditions.

a) The MDEV values obtained from the normative model (labeled as "N") were larger than the mean MDEV values obtained from the adaptive model (labeled as "A") for any given run condition. This is expected because in the computation of differences in the lateral positions, the adaptive model uses the subject's actual lateral profile obtained while performing the primary task only during the qualifying run after the practice trials as the reference profile. Whereas the normative model uses the ideal lateral profile as the reference to compute the differences in lateral position.

b) The baseline values (labeled as "Base 1" and "Base 2") of MDEV were about 0.7 m and 0.35 m, respectively, for the normative and the adaptive models.

c) The MDEV values of all other runs with the dual tasks were larger than the corresponding baseline values obtained under the primary (single) task in baselines 1 and 2.

d) The mean MDEV value for "easy" level of any given secondary task was smaller than the mean MDEV value of the "hard" level of the secondary task. Thus, the MDEV measure was found to be sensitive to differences in the levels of the secondary tasks.

e) The magnitude of the difference in the mean MDEV values in any given dual task run and the MDEV value in the corresponding mean MDEV value of the baseline, thus, indicates (or measures) the amount of additional workload imposed by the secondary task.

CONCLUDING COMMENTS

The above described studies show that currently many methods and performance measures are used to determine the effects of addition of different in-vehicle tasks to the baseline driving situations. The NASA TLX, SWAT, WP, SAE's total task time and ISO's LCT are examples of methods, each of which has the capability to provide a single overall measure of driver workload. However, criterion limits are presently not established to determine acceptable and unacceptable levels of driver workload and each method has some major shortcomings. The subjective methods cannot be used without the availability of actual prototype hardware; and extensive time is required to conduct evaluations by using actual subjects. The LCT test procedure lacks the validity because the drivers in the real world driving do not change lanes continuously at the high frequency rate built into the procedure; and most drivers under most driving situations can voluntarily decide on when and how quickly to change a lane. The SAE J2364 and 2365 test procedures are not based on actual driving.

While many of the above-discussed methods provide useful information, none of them can be used alone to decide on the acceptability of any new in-vehicle feature. The industry would prefer to use methods that are objective, less time consuming, repeatable and precise. Thus, until better methods are developed, the decision makers will have to continue to rely on using combinations of many of the existing workload measurement methods; and supplement their findings with additional information obtained from other sources, such as ergonomics experts, customers and benchmarking prototypes with other available in-vehicle devices.

REFERENCES

Bhise, V. D. and S. Bhardwaj. 2008. *Comparison of Driver Behavior and Performance in Two Driving Simulators*. SAE Paper# 2008-01-0562, Presented at the SAE World Congress, Detroit, MI.

Brookhuis, K. A. and D. Waard. 2000. Assessment of Driver's Workload: Performance and Subjective and Physiological Indexes. In *Stress, Workload and Fatigue*, ed. P. A. Hancock and P. A. Desmond, 321–333. Boca Raton, FL: CRC Press.

Card, S. K., T. P. Morgan, and A. Newell. 1980. The Keystroke-Level Model for User Performance Time with Interactive Systems, *Communications of the ACM*, 23(7): 396–410.

Card, S. K., T. P. Morgan and A. Newell. 1983. *The Psychology of Human-Computer Interaction*. Hillsdale, NJ: Lawrence Erlbaum Associates.

Cooper, J. M. and D. L. Strayer. 2008. Effects of Simulator Practice and Real-World Experience on Cell Phone Related Driver Distraction. *Human Factors*, 50(6): 893–902.

Drews, F., H. Yazdani, C. N. Godfrey, J. M. Cooper and D. L. Strayer. 2009. Text Messaging During Simulated Driving. *Human Factors*, 51(5): 762–770. Also published in *Journal of the Human Factors and Ergonomics Society*, Online First, published on December 16, 2009 as doi:10.1177/0018720809353319 (9 pages).

Hankey, J. M., T. A. Dingus, R. J. Hanowski, W. W. Wierwille and C. Anderws. 2001. In-Vehicle Information Systems Behavioral Model and Design Support: Final Report. Report No. FHWA-RD-00-135 sponsored by the Turner-Fairbank Highway Research Center of the Federal Highway Administration, Virginia Tech Transportation Institute, Blacksburg, VI.

Hart, S. G. and L. E. Staveland. 1988. Development of NASA-TLX (Task Load Index): Results of Empirical and Theoretical Research. In *Human Mental Workload*, ed. P. A. Hancock and N. Meshkati, 139–183. Amsterdam: North Holland.

Hitt II, J. M., J. P. Kring, E. Daskarolis, C. Morris and M. Mouloua. 1999. Assessing mental workload with subjective measures: An analytical review of the NASA-TLX index since its inception.

Hosking, S. G., K. L. Young and M. A. Regan. 2009. The Effects of Text Messaging on Young Drivers. *Human Factors*, 51: 582–592.

International Standards Organization (ISO). 2008. Road Vehicles – Ergonomic Aspects of Transportation and Control Systems – Simulate Lane Change Test to Assess In-Vehicle Secondary Task Demand. Draft of ISO/DIS 26022, prepared by the Technical Committee ISO/TC22, Road Vehicles, Subcommittee SC13, Ergonomics.

Jackson, D. and V. D. Bhise. 2002. *An Evaluation of the IVIS-DEMAnD Driver Attention Model.* SAE Paper no. 2002-01-0092, Paper presented at the SAE International Congress in Detroit, MI.

Jothi, V. 2009. Applications of the ISO Lane Change Procedure. Master's Thesis, the University of Michigan-Dearborn, Dearborn, MI.

Lambert, S., S. Rollins, and V. D. Bhise. 2005. Effects of Driver Induced Distraction Tasks on Driver Performance and Glance Behavior. In *Proceedings of the Annual Meeting*, Transportation Research Board, Washington, D.C.

Mehler, B. R., B. Reimer, J. F. Coughlin and J. A. Dusek. 2009. Impact of Incremental Increases in Cognitive Workload on Physiological Arousal and Performance in Young Adult Drivers. Transportation Research Record, *Journal of the Transportation Research Board*, No. 2138.

Meshkati, N., P. Hancock and M. Rahimi. 1992. Techniques in Mental Workload Assessment. In *Evaluation of Human Work, A practical Ergonomics Methodology*, ed. J. Wilson and E. Corlett, 605–627, London: Taylor and Francis.

Olsson, S. and P. C. Burns. 2000. *Measuring Driver Visual Distraction with a Peripheral Detection Task.* Sweden: Department of Education and Psychology, Linkoping University.

Owens, J. M., S. B. McLaughlin and J. Sudweeks. 2010. On-Road Comparison of Driving Performance Measures When Using Handheld and Voice-Control Interfaces for Cell Phones and MP3 Players. SAE Paper no. 2010-01-1036, Presented at the 2010 SAE World Congress held in Detroit, MI.

Reid, G. B. and T. E. Nygren. 1988. The Subjective Workload Assessment Technique: A scaling procedure for Measuring Mental Workload. .In *Human Mental Workload*, ed. P. A. Hancock and N. Meshkati, 139–183. Amsterdam: North Holland.

Rubia, S., E. Diaz, J. Martin and J. M. Puente. 2004. Evaluation of Subjective Mental Workload: A Comparison of SWAT, NASA-TLX, and Workload Profile Methods. *Applied Psychology: AN International Review*, 53(1): 61–86.

Society of Automotive Engineers, Inc. 2009a. SAE J2364 Recommended Practice: Navigation and Route Guidance Function Accessibility While Driving. In *SAE Handbook,* Published by the Society of Automotive Engineers, Warrendale, PA.

Society of Automotive Engineers, Inc. 2009b. SAE J2365 Recommended Practice: Calculation of the Time to Complete In-Vehicle Navigation and Route Guidance Tasks. In *SAE Handbook*, Published by the Society of Automotive Engineers, Inc, Warrendale, PA.

Tijerina, L., E. Parmer and M. J. Goodman. 1999. Driver Workload Assessment of Route Guidance System Destination Entry While Driving: A Test Track Study. Transportation Research Center, East Liberty, Ohio.

Tijerina et al. 2000. Driver Distraction with Wireless Telecommunications and Route Guidance Systems. Report no. DOT HS809-069, National Highway Traffic Safety Administration, Washington, D.C.

Tsang, P. S. and V. L. Velazquez. 1996. Diagnosticity and Multidimensional Subjective Workload Ratings. *Ergonomics*, 39(3): 358–381.

Verwey, W. B. and D. M. Zaidel, 2000. Predicting Drowsiness Accidents from Personal Attributes, Eye Blinks, and Ongoing Driving Behaviour. *Personality and Individual Differences*, 28(1): 123–142.

Wickens, C. D. 1992. *Engineering Psychology and Human Performance*. New York, NY: HarperCollins.

Wierwille, W.W. and F. T. Eggemeier. 1993. Recommendation for mental workload measurement in a test and evaluation. *Human Factors,* 35(2), 263–281.

17 Ergonomic Considerations in Electric Vehicle Development

INTRODUCTION

WHAT IS AN ELECTRIC VEHICLE?

An electric vehicle (EV) is driven purely on electric power. An EV can operate using one or more electric motors for propulsion instead of an internal-combustion engine (ICE) that generates power by burning a mix of fuel and gases. The electric energy required to run an EV can be obtained from a) energy stored in batteries, b) electric power generated by on-board sources (e.g., hydrogen fuel cells, solar panels, or a generator operated by an ICE), and c) connecting it to a wiring system (e.g., through an overhead pantograph or induction devices embedded in the pavement). EVs are not limited to road vehicles (automotive products) but can also include electric motor operated rail vehicles, surface and underwater vessels, electric aircraft and electric spacecraft. EVs are called the battery electric vehicles (BEV) when they also carry the battery required to generate the electric energy used to run their electric motors.

Future advances in abilities to a) increase energy storage capabilities, b) reduce battery weight, volume and cost, and c) increase in charging stations will increase driving distance (range), and thus, accelerate their market share.

System Architecture of Electric Automotive Products

There are a number of variations in how the electric vehicle is configured and its power is delivered. These variations are described below.

1. *Battery Electric Vehicles* (BEVs): BEVs are also known as All-Electric Vehicles (AEV).
2. *Hybrid Electric Vehicle* (HEV): The HEV includes an internal combustion engine along with one or more electric motors. The hybrid powertrain consumes less fuel because a portion of its input energy is supplied by the electric motor, and it is more efficient than the internal combustion engine. Further, during vehicle deceleration the electric motor acts like a generator and recovers dynamic energy of the vehicle and uses it to recharge the battery.

DOI: 10.1201/9781003485605-4

There are different ways that a hybrid electric vehicle can combine the power from one or more electric motors and the internal combustion engine. Electric motors can be attached to the IC engine in parallel or series configuration.

Parallel Hybrid: The most common type is a parallel hybrid that connects the engine and the electric motor to the wheels through mechanical coupling. In this configuration, both the internal combustion engine and the electric motor provide power to the drive wheels.

Series Hybrid: In this configuration, the drive wheels are powered by an electric motor and the internal combustion engine drives an alternator which charges the battery. The electric motor is driven by the battery through an electronic module. This configuration is often referred to as extended-range electric vehicles (EREVs) or range-extended electric vehicles (REEVs).

Series-Parallel Hybrid: There are also series-parallel hybrids where the vehicle can be powered by the engine working alone, the electric motor on its own, or by both working together; this is designed so that the vehicle can run at its optimum range as often as possible.

3. *Plug-in electric vehicle* (PEV): The PEV is any motor vehicle that can be recharged from any external source of electricity, such as a wall socket. The electricity stored in the rechargeable battery packs drives or contributes to drive the wheels. PEV is a subcategory of electric vehicles that includes battery electric vehicles (BEVs), plug-in hybrid vehicles (PHEVs), and electric vehicle conversions of hybrid electric vehicles and conventional internal combustion engine vehicles.

4. *Two or More Motors*: Some EVs and hybrid powertrains have two or more electric motors (e.g., each wheel motor directly drives a wheel or front and rear motors drive the front and rear wheels, respectively).

SPECIFICATIONS FOR AN ELECTRIC CAR: A CASE STUDY

OBJECTIVE

The objective of this project was to illustrate the use of the matrix data analysis technique to develop specifications of an electric vehicle.

PROJECT BACKGROUND

A research project was undertaken in the author's automotive engineering course to develop specifications for a future electric car. The matrix data analysis technique was used to translate the customer needs of an electric vehicle into its specifications. The matrix data analysis technique is used in the Quality Function Deployment (QFD) to translate the customer requirements into the functional specifications of a product (see Volume 1, Chapter 3 for details on the QFD) and the matrix data analysis in (Bhise, 2023; Besterfield et al., 2003).

APPLICATION OF MATRIX DATA ANALYSIS

The following customer requirements were identified by interviews with potential drivers of future electric cars:

1. Long travel range on one-time charge
2. Short time to fully recharge
3. Greater electric energy efficiency
4. Hassle-free charging from home or office
5. Sound feedback on vehicle movement provided by the EV to the driver and other nearby persons
6. Safety system to prevent short circuits, current leakage, and fires
7. Vehicle parameters monitoring (e.g., speed, temperatures of the motor and battery)
8. Simple, easy to use, and informative interfaces
9. Constant display for information (e.g., range display, energy use display [kwh/km or kwh/mile], cost display [cents/km or cents/mile], and battery charge gauge [percentage of battery capacity available])
10. Memory storage space/capacity for trip data, songs, phonebook, and so forth
11. Availability of trip data and analysis of vehicle information from a remotely located PC
12. Handheld device (or an interface to a smart phone) to carry vehicle information and help in trip planning

The areas for engineering specifications included in the analysis were as follows:

1. Type of battery
2. Charge of the battery
3. Buffer stock of charge
4. Torque of the motor
5. Power of the motor
6. Driver-vehicle (driver-to-vehicle) interface
7. House-utility company interface
8. Utility company–vehicle (utility company-to-vehicle) interface
9. Vehicle–house (vehicle-to-house) interface
10. Data connectivity
11. Power connectivity
12. Outside safety (i.e., safety of people outside the vehicle due to electrical hazards and sound level to hear an approaching electric vehicle)
13. Inside safety (safety of people inside the vehicle due to electrical hazards)

The matrix data analysis chart presented in Table 17.1 was completed by the researchers by determining the importance ratings of the customer requirements (second column from left) and the strength of relationships between the customer requirements and the functional specifications (the matrix of 12 rows and 13 columns) based on the information obtained from the driver interview data and literature surveys conducted for this project. The importance of each customer need was

TABLE 17.1
Matrix Data Analysis Chart for Electric Car

Column Number			1	2	3	4	5
Row Number	Importance Rating	Description of Customer Need	Type of battery	Charge of the battery	Buffer stock of charge	Torque of the motor	Power of the motor
1	10	Long travel range on one-time charge	9	9	9	3	3
2	9	Less time to fully recharge	9	9	9	1	1
3	7	Electric energy efficiency	9	1	1	9	9
4	9	Hassel-free charging from home or office	9	1	1	1	1
5	8	Sound feedback	1	1	1	9	9
6	9	Safety system to prevent short-circuits,leakage, fires, etc.	3	3	1	1	1
7	7	Vehicle parameters monitoring	3	9	9	9	9
8	9	Simple easy to use and informative interfaces	1	1	1	1	1
9	8	Constant display for information (e.g., kwh/mile)	9	9	9	1	1
10	6	Memory storage space for trip data, songs, etc.	1	1	1	1	1
11	6	Availability of trip data and analysis from a remote PC	1	1	1	1	1
12	7	Handheld device to carry vehicle info. & trip planning	1	3	1	1	1
		Absolute Importance Ratings	471	399	367	291	291
		Relative Importance Ratings	8.8	7.4	6.8	5.4	5.4

6	7	8	9	10	11	12	13
Driver-vehicle interface	House-utility company interface	Utility company to vehicle interface	Vehicle-house interface	Data connectivity	Power connectivity	Outside safety	Inside Safety
9	1	3	1	1	1	1	1
1	9	9	9	1	3	1	1
9	9	3	1	3	9	1	1
9	3	1	9	1	9	9	9
9	1	1	1	1	1	9	1
9	1	1	1	3	9	9	9
9	1	1	9	9	9	9	9
9	1	1	9	1	1	9	9
9	1	1	9	9	9	1	1
9	1	3	3	1	3	3	3
9	1	3	3	9	1	1	1
9	3	9	9	9	1	9	9
783	255	281	511	351	445	499	435
14.6	4.7	5.2	9.5	6.5	8.3	9.3	8.1

rated by using a 10-point scale, where 10 = most important and 1 = least important (see second column with heading "Importance Rating" in Table 17.1). Each cell of the relationship matrix provides a number to illustrate the strength of relationship between the customer need and functional specification defining the cell. The weights of 9, 3, and 1 were used to define strong, medium, and low relationships, respectively.

The bottom two rows of the chart show the absolute importance ratings and relative importance ratings of the specifications (see Volume 1, Chapter 3 for calculations of these ratings in the QFD section). The relative importance weight row at the bottom of the matrix data chart shows that the three most important engineering requirements (specifications) are as follows: (1) driver–vehicle interface (14.6% relative importance rating); (2) vehicle–house interface (9.5% relative importance rating); and (3) outside safety (9.3% relative importance rating).

Advantages and Disadvantages of Electric Vehicles

This section provides brief descriptions of advantages and disadvantages of electric vehicles to help the reader in understanding issues related to electric vehicles.

Advantages

1. *Convenience*: The vehicle can be charged at home. Thus, there is no need to find and go to a gas/charging station for recharging. Electric vehicles can be charged using wired plug-in chargers, which provide more flexibility as compared to the gas stations used for refueling ICE equipped vehicles. They can be charged at home, in charging stations at work, in public, and on highways for long-distance trips. They can also be charged by mobile charging vehicles (which carry a generator for charging). People tend to follow a charging hierarchy that starts at home. Currently, they are good for short-distance driving and charging from home.

2. *Reduced energy cost*: Electric vehicles are cheaper to use as compared to the ICE-equipped vehicles. They typically cost about 2–3 cents/mile as compared to about 10–15 cents/mile for gasoline-powered vehicles. (Note that currently [in 2023] the consumers pay on average about 19 cents/kWh of electricity in the United States. However, many recharging stations have been charging about 49 cents/kWh.). Regenerative braking used in EVs recovers energy used in braking. Thus, electric vehicles are more energy-efficient than the ICE equipped vehicles. Further, most individual passenger cars remain parked for 8 to 12 hours at night, and home charging can be easy and often cheaper than charging elsewhere. Also, most charging can happen overnight when off-peak electricity prices are lower.

3. *Cleaner non-polluting*: EVs do not generate exhaust gases. Thus, they will reduce environmental pollution and resulting health related (e.g., respiratory, and cardio-vascular) problems. More renewable electric sources can be added to the grid as the power demand on the grid will increase. Also, recycling of

lithium used in the batteries is expected to reduce pollution generated during extraction, processing of lithium and battery production in future.

4. *Low sound*: EVs produce much lower sound levels than vehicles with their exhaust systems. EV's do not produce sound when stopped. Thus, EVs are quieter under idling situation than ICE equipped vehicles (except with stop-start capability).

5. Higher acceleration: Newer electric vehicles have more powerful electric motors that can provide higher accelerations than most gasoline powered vehicles. Electric motors produce higher torque output at low motor speeds as compared to the ICE equipped vehicle. Thus, EVs accelerate faster and are fun-to-drive.

6. *Package efficient*: EVs do not need transmissions and exhaust systems. The electric motor in an EV is much smaller (in volume) than an ICE of similar power. It can be placed right between the driven wheels and a few companies have built EVs with motors inside the wheel hubs. An EV needs no transmission as such, so there is no need for the central tunnel that takes up so much space in rear-wheel-drive ICE vehicles. There is also no need for an exhaust system, thermal shielding, or a catalytic converter. EV batteries are typically packaged under the vehicle floor between the front and rear wheels. This battery location also reduces the center of gravity of the vehicle which also improves vehicle stability and handling performance.

7. *Cheaper batteries*: The prices of batteries are also more likely to decrease with technical advances. The increased availability of cheaper batteries will also increase sales of EVs and the use of the batteries to store excess energy produced from renewable energy sources.

8. *Incentives*: The Inflation Reduction Act currently provides generous credits for battery production, and EV buyers are modestly rewarded with rebates up to $7,500, thus further easing the costs of this government-forced transition from ICE equipped vehicles to EVs.

9. *Back-up Home/Service Generators*: Some of the currently available EVs can be used to power homes during power failures. Electric pickup trucks can also provide power needed to run tools and equipment used at construction/service sites.

Disadvantages

1. *Range anxiety*: Range anxiety is a very frequently mentioned problem with EVs because the drivers may not be able to find a battery charging station when needed. Further the estimate of available range (i.e., the distance the vehicle can travel before the battery power is exhausted) is affected by many variables such as changes in vehicle velocity, temperatures, air-conditioning and heating demands and other electrical equipment in the vehicle.

2. *Longer recharging time*: The battery recharging time for an EV is longer than refueling time for an ICE-equipped vehicle.

3. *Too few recharging stations*: Recharging stations for EVs are currently not as easily available as gas stations. This results in "range anxiety" for many drivers.

4. *Need for recharging infrastructure*: Charging infrastructure is currently not sufficiently developed. Infrastructure also needs to be standardized for customer convenience. Finding charging stations is difficult. Charging costs at remote charging stations would most likely be much higher than charging at home. Approximately 3 to 6 percent of total miles driven involve long-distance trips that average more than 100 miles. Even with a full charge leaving home, most of today's EVs cannot make that round-trip without recharging. This makes the case for long-distance chargers. Drivers without chargers at home or work must charge in public; drivers who exceed their battery range on a given day may need to visit fast-charge stations; and drivers who forget to charge at home or who do not have home chargers must rely on other options, making the case for public charging.

5. *Cabling inconvenience*: Process of charging by plugging the cable into the vehicle and unplugging the cable and storage of cable is inconvenient. In future, wireless charging would eliminate the hassle of handling and plugging cables. Currently available electric vehicles also differ in terms of location of the electricity input port (e.g., left or right side of the vehicle, front, middle or rear of the vehicle) and the type of the port (e.g., Tesla designed plug, or a special adaptor is required). Further, variations also exist in the software applications of the vehicle manufacturers used in their cell phones or vehicles to control the charging and payment process. These variations have caused additional driver confusion and frustrations in the charging process.

6. *Battery weight and size*: The battery in EV weighs the most as compared to other systems in the vehicle. The battery volume is also large. Thus, batteries are typically packaged under the vehicle floor between the front and rear wheels.

7. *Reduced battery output at low and high temperatures*: The battery output reduces as the ambient (battery operating) temperature decreases. Thus, the batteries provide lower range under low ambient temperatures. Further, greater use of electric heaters and air circulating fans in colder weather conditions consume additional electric power. Under higher ambient temperatures (e.g., summertime in southwestern states), air-conditioning systems are used heavily. The constant use of air-conditioning systems will also substantially reduce the range of electric vehicles.

8. *Low energy storage capacity*: Battery storage capacity (energy density measured in kWh/kg or kWh/l) is low to run vehicles over longer distances. The energy density of gasoline is much higher than currently available lithium-ion batteries. Large capacity batteries will be heavy, and they will require larger electric motors to drive the heavier vehicles. Battery advancement may take many more years. Further use of light-weight materials to reduce vehicle weight is costly in comparison to steel.

9. *Power loss*: The EVs will lose about 30% of electric power in distributing and charging, and in the losses within the vehicle electric systems (e.g., losses in electric drive system, maintaining temperatures in batteries, and running power steering and climate control). Thus, about 30% of the input power used to charge an EV will be lost. (Note: EVs convert over 77% of the electrical energy from the grid to power at the wheels. Conventional gasoline vehicles only convert about 12%–30% of the energy stored in gasoline to power at the wheels [DOE, 2023]).

10. *Cost of the EV*: EVs cost much more (about $55,000) as compared to ICE engine-equipped vehicles (about $35,000). (Currently, the automakers' enormous investments in EVs appear to be largely driven by political choices related to incentives in vehicle prices and meeting future fuel economy and emissions requirements, and not by the consumer's choice.)

11. *Future electricity costs*: The costs related to the electrification of the automotive market are likely to increase in the future as more EVs are sold. The cost of electricity will also increase as more features of the smart grid (e.g., digital two-way communication between the utility company and the EV, and circuits monitoring and automatically controlling power switching capabilities) are incorporated.

12. *Additional sources of electric power*: Currently, most electric generating plants (that provide battery recharging power) run on coal or natural gas, which emit greenhouse gases. The costs of hooking up of many smaller renewable energy power plants (e.g., solar and wind) to the existing grid and maintenance costs of the power grid are expected to increase the electricity power transmission and distribution costs in the future. These new renewable power sources can only provide power on an interruptible basis (solar during daytime and wind power at high wind speeds) and will require back-up power through battery storage or fossil fuel power plants.

ERGONOMIC CONSIDERATIONS IN DESIGNING AN EV

The uses of an EV and the users' tasks in operating an EV are not very different from the traditional ICE equipped vehicle except for the recharging of the batteries. Thus, the ergonomic considerations in designing an EV are as follows:

1. Design considerations related to occupant packaging (e.g., positioning of occupants and seats, locations of controls and displays, field of view, exterior lamps, and entry-exit) should be the same as the ICE equipped vehicles. The EVs should be designed to meet the same SAE occupant packaging practices and the FMVSS 100 series standards as the ICE equipped vehicles (SAE, 2023; NHTSA, 2023).

2. The occupant protection in an EV should meet the same impact protection requirements as specified in the FMVSS 200 and 300 series standards (NHTSA, 2022). Thus, the design of seats, steering column, air bags, seat

belts and roof pillars are expected to have the same impact protection and ergonomic design considerations.

3. The driver should be provided information about the state of vehicle electrical system–especially parameters that display readiness of electrical system, battery charge, miles to recharge, and regenerative braking system.

4. The driver should be provided with the displays and controls to find and select EV functions quickly when EV operating characteristics need to be reset or changed (e.g., reduction in automated climate control usage when battery charge is low, or when the battery temperature increases rapidly).

5. Because of their electric drivetrain and large heavy battery, EVs may provide a different driving feel in comparison to the ICE equipped vehicles. Additional information related to these differences should be provided in the owner's manual. Higher weight of the EV (due to battery weight) and larger torque at lower speeds in comparison to an ICE equipped vehicle will affect maneuvering capabilities under certain situations (e.g., vehicles requiring higher ground clearance and off-road driving).

6. For reducing the driver's range anxiety by displaying accurate information on miles to recharge is important as additional electric load is added or reduced by auxiliary or on-board equipment.

7. Loading/unloading from frunk. Reduce height of load lift-off to load (from the ground) or unload items in the frunk. (Hood should lift the grill area to reduce lifting over the grill height). Provide low lift-off and low load floor.

8. Electric vehicles will have more electrical hardware than ICE-equipped vehicles to control many electrical systems. These electrical systems will involve extensive software codes and applications. Problems in seamless integration and updating of the software need to be managed by the auto companies to avoid user inconveniences and frustrations in using the vehicle systems (e.g., accessing electrical charging system and recharging payment management system using cell phones).

9. EVs can serve as a back-up electricity generator for homes or at other locations (e.g., construction sites). Additional display functions should be provided in the vehicle to set and manage the vehicle generator and battery power output, discharge rate and warning messages when the time to recharge reaches a low level.

INFRASTRUCTURE STANDARDIZATION FOR ELECTRIC VEHICLES

To accommodate the electric vehicles (EVs) within the nation's existing highway transportation and economic system, the infrastructure must be expanded to serve the needs of the growing EV users. The needs of the customers are to allow them to a) recharge their electric vehicles quickly, economically and safely, b) provide additional convenience and comfort features and facilities offered by existing gas stations and highway plazas such as restrooms, food shops, restaurants, entertainment options and ATM machines. Or the charging stations could be integrated with the existing facilities that provide fuel filling and other convenience features. Ergonomics engineers should be involved in developing the needed infrastructure and its features to meet the needs of the EV users as well as users of other types of vehicles.

What Is Infrastructure for Electric Vehicles?

The term infrastructure normally includes all the systems and structures, including access roads and utilities required to provide the required service (with all necessary functions) to the area it supports. For example, the electrical grid across a city, state or country is the infrastructure based on the power generation, transmission and distribution equipment and supporting facilities to meet electricity demand to the areas it supports. Similarly, the physical structures (e.g., buildings), wiring, cabling, wi-fi communications and components making up the data network operating within a specific location are also parts of the infrastructure for a business. Thus, the infrastructure can be classified into the following three categories:

1) *Hard Infrastructure*: These make up the physical systems that make it necessary to run a modern, industrialized nation. Examples of physical systems for EV-based transportation here will include vehicle charging equipment, electric power transmission and generating capabilities, data communication and processing equipment, associated buildings, roads, bridges, as well as the capital/assets needed to make them operational and user-friendly.

2) *Soft infrastructure*: These types of infrastructure make up institutions that help maintain the economy. These usually require jobs and people to provide and maintain the required services.

3) *Critical Infrastructure*: These are assets defined by a government as being essential to the functioning of a society and economy, such as safe transportation and facilities to support related functions such as surveillance and policing, shelter and heating, telecommunication, and public health.

From the viewpoint of convenience of users of the EVs, customers should have similar, or even more advanced, facilities and services available when they use their EVs as they use their conventional internal combustion engine (ICE) powered vehicles. The charging ports for the electric vehicle will be different from the fuel filling ports. Standardizing charging interfaces between electric vehicles and charging stations is one of the topics covered in the infrastructure standardization. The equipment should be easy to use, that is, not require many different types of interfaces, adapters, and other equipment/systems. The equipment and processes involved in charging and billing should be easy to learn and use. Overall, the whole process of charging should thus reduce user difficulties, frustrations and errors, and keep the charging time and costs as low as possible. Other issues related to standardization are transaction processing, for example, billing and payments for the services.

As the EV industry matures, interoperability (i.e., open communication and exchange of data; see next section) will continue to remain important to the development of vehicle hook-ups and communications with the chargers, the grid, and the power company. We can expect to see further advancements in the communication systems between electric vehicles and grid-connected assets. Intelligent power supply is an emerging technical and commercial opportunity carrying many benefits across the vehicle electrification landscape and will undoubtedly demand open and harmonized communication standards (Bablo, 2016).

While EV infrastructure is still a relatively new and quickly evolving space, regardless of which vehicles and charging innovations will capture and drive the market, open standardization will always be the optimal approach for rolling out the most future-proof and reliable charging infrastructure. The interoperability strategy can deliver the most returns for those who will fund, deploy, operate, and use these critical assets in the years to come – for the most convenient, reliable, and clean transportation.

As more electric vehicles replace the ICE-powered vehicles, the demand for electricity will also increase. Additional electricity generating capacity along with rules and incentives to redistribute the available electric supply (e.g., by charging EV's at off-peak and nighttime hours at lower rates when electricity demand from other equipment is less) are possible solutions always under consideration (Gold, 2020). When too many drivers want to fast-charge their EVs, the load on the grid can increase substantially. And the power grid may need upgrading with additional power distribution and power generation capabilities. For example, since most people who own EVs usually charge them at home, it would mean changes in substations and distribution circuits to accommodate multiple homes in a neighborhood drawing power to recharge the EV batteries.

The issues related to the infrastructure are:

Physical infrastructure:

a) Number of charging stations
b) Parking spaces for vehicles to be charged should include waiting areas for vehicles. The number of parking spaces will depend upon the time required for full charge and traffic flow in the road network near the charging station.
c) Electrical grid (with available sources, their capacity, distribution network and characteristics of power demands) in the charging area and in the vicinity of the charging area. The capacity of the grid in kWh to provide electric power required to charge all vehicles at the charging stations. Back-up electrical sources (e.g., standby generators) and emergency electrical storage systems in case of grid failures.
d) Restaurants and comfort stations with communication capabilities (e.g., telephone, wi-fi routers). The comfort stations should meet the needs of drivers requiring longer charging hours for the EVs.
e) Access to tow trucks, service vehicles for disabled vehicles, firefighting equipment, and emergency vehicles

Critical Infrastructure:

a) Local management personnel for service, safety, shops, and restaurants personnel

b) Level of automation involved in minimizing charging time and communicating charging progress with customers

Soft Infrastructure:

a) Number of fulltime jobs by each needed job classification (e.g., maintenance personnel, electrical and computer, safety, and security personnel) dependent upon the charging facility
b) Incentives from local, state, and federal governments
c) Incorporation of customer convenience and satisfaction features with the whole charging experience (e.g., ability to reserve a charging spot within a specified time interval)

What Is Interoperability?

Interoperability, in the most universal terms, is the open communication and exchange of data between and among devices and/or software systems. Interoperability is a key issue for many industries such as software development, home automation, healthcare, telecommunications, and public safety. Interoperability has many benefits, such as varied mobile devices to work across different cellular networks in different regions; or when communities rely on police and fire departments to communicate with each other using common platforms during emergencies.

The term is often used to describe multiple aspects of electric vehicle charging, and can include form factor, communication, and compatible ratings among any of the following entities in a charging system:

a) The vehicle
b) The charging station hardware for conductive as well as the wireless power transfer technologies and battery swapping capabilities
c) The charging station connectivity software
d) The back-office or payment back-end
e) The network operators
f) The energy management system
g) The power supply

CHARGING STATIONS AND CHARGERS FOR ELECTRIC VEHICLES

Charging Stations

There are essentially 3 types of charge stations that are widely in use, but they are designated differently around the globe.

In North America, there is the Portable EV Cord Set, a portable device intended to stay with the vehicle and be used with any convenient receptacle. In the IEC document, this was referred to as a Mode 2 cable assembly (IEC, 2017).

Second, in North America there is a fastened in place charge station, which is a device that can be moved but is not intended to be moved often. It typically is "hung on a hook" in a residential or commercial garage for use with electric vehicles that are parked in the vicinity. In the IEC document, this is a Mode 2 charge station.

The last device in North America is a fixed charge station, typically a public access charge station that is permanently fixed in one location and is hard wired. In the IEC document, this is designated as a Mode 3 charge station. The designations are not all that important to the discussion except for the portable EV cord set / Mode 2 cable assembly.

TYPES OF CHARGERS

To get the most out of a plug-in electric vehicle, it must be charged on a regular basis. Charging frequently maximizes the range of all-electric vehicles and the electric-only miles of plug-in hybrid electric vehicles. Drivers can charge at home, at work, or in public places. While most drivers do more than 80% of their charging at home and it is often the least expensive option, workplace and public charging can complement.

Conductive Charging

Conductive charging can be accomplished by delivering AC power to a vehicle with an on-board charger or by delivering DC power to a vehicle for directly charging the battery (no on-board charger needed). In some cases, the vehicle may contain both an AC and DC connection, and in such cases, the on-board charger is used, as necessary. AC delivery is done through what will be called a charging station. DC delivery is done through what will be called a quick charger. Charge stations and quick chargers, although different internally and perhaps using different communication protocols, all must be provided with a means to connect conductively to the vehicle. This is done through an output cable and an EV connector. Standards covering these products are also included in the discussion within the SAE and IEC committees. A system of protection for the user of the charging system is also provided internally to the charge station or the quick charger. It is designated and treated differently by different standards.

Plugging into a Charger

Charging an EV requires plugging into a charger connected to the electric grid, also called electric vehicle supply equipment (EVSE). There are three major categories of chargers, based on the maximum amount of power the charger provides to the battery from the grid (DOE, 2020c). They are SAE level 1, level 2, and level 3 chargers (SAE J1772, SAE, 2023).

a) *Alternate-current (AC) charging, also known as level 1 or level 2.*

In this system, an in-car inverter converts AC to direct current (DC), which then charges the battery at either level 1 (equivalent to a US household outlet) or level 2 (240 volts). It operates at powers up to roughly 20 kW.

Level 1: Provides charging through a 120 V AC plug and does not require installation of additional charging equipment. Can deliver 2 to 5 miles of range per hour of charging. Most often used in homes, but sometimes used at workplaces.

Level 2: Provides charging through a 240 V (for residential) or 208 V (for commercial) plug and requires installation of additional charging equipment. It can deliver 10 to 20 miles of range per hour of charging and it is used in homes, workplaces, and for public charging.

The kilowatt capacity of a charger determines the speed at which the battery receives electricity. AC level 1 and level 2 are most applicable for homes and workplaces because of the long parking (recharging) periods and their lower cost: a simple level 2 for a home can cost as little as $500.

Basic AC level 1 and level 2 power will overwhelmingly remain the dominant charging technology through 2030, providing from 60 to 80 percent of the energy consumed. Most of this charging will take place at homes, workplaces, and via slow-charge public stations.

b) *DC charging*

This charging system converts the AC from the grid to DC before it enters the car and charges the battery without the need for an inverter. Usually called direct-current fast charging or level 3, it operates at powers from 25 kW to more than 350 kW.

DC Fast Charge: This provides charging through 480 V AC input and requires highly specialized, high-powered equipment as well as special equipment in the vehicle itself. (Plug-in hybrid electric vehicles typically do not have fast charging capabilities.) It can deliver 60 to 80 miles of range in 20 minutes of charging. It is used most often in public charging stations, especially along heavy traffic corridors.

Charging times range from less than 30 minutes to 20 hours or more based on the type of EVSE, as well as the type of battery, how depleted it is, and its capacity. All-electric vehicles typically have more battery capacity than plug-in hybrid electric vehicles, so charging a fully depleted all-electric vehicle takes longer. Direct Current Fast Charging (DCFC) chargers are most applicable in situations where time matters, such as on highways and for fast public charging. DCFC will likely play a much larger role in China, which requires more public-charging infrastructure.

c) *Types of Plugs*

Most modern chargers and vehicles have a standard connector and receptacle, called the SAE J1772 connector (SAE, 2023). Any vehicle with this plug receptacle can use any level 1 or level 2 EVSE. All major vehicle and charging system manufacturers support this standard, so the EV should be compatible with nearly all non-fast charging workplace and public chargers.

Fast charging currently does not have a consistent standard connector. The SAE International, an engineering standards-setting organization, has passed a standard for fast charging that adds high-voltage DC power contact pins to the SAE J1772 connector currently used for level 1 and level 2. This connector enables use of the same receptacle for all levels of charging and is available on certain models like the Chevrolet Spark EV. However, other

EVs (the Nissan Leaf and Mitsubishi i-MiEV in particular) use a different type of fast-charge connector called CHAdeMO. Fortunately, an increasing number of fast chargers have outlets for both SAE and CHAdeMO fast charging. Lastly, Tesla's Supercharger system can only be used by Tesla vehicles and is not compatible with vehicles from any other manufacturer. Tesla vehicles can use CHAdeMO connectors through a vehicle adapter.

Wireless Charging

In addition to the three types above, wireless charging uses an electro-magnetic field to transfer electricity to an EV without a cord. This system uses electromagnetic waves to charge batteries. There is usually a charging pad connected to a wall socket and a plate attached to the vehicle. Current technologies align with level 2 chargers and can provide power up to 11 kW. The Department of Energy is supporting research to develop and improve wireless charging technology. Wireless chargers are currently available for use with certain vehicle models.

SAE International recently published two new documents, SAE J2954 and SAE J2847/6, which ensure a safe and efficient method for transferring power from charging stations to electric vehicles. SAE J2954 Standard: "Wireless Power Transfer and Alignment for Light Duty Vehicles" establishes the first standard for wireless power transfer (WPT) for both electric vehicle and electric vehicle supply equipment (EVSE). This enables light-duty electric vehicles and infrastructure to safely charge up to 11 kW, over an air gap of 10 inches (250 mm), achieving up to 94 precent efficiency.

The SAE J2954 standard is a game-changer by giving a "cookbook" specification for developing both the vehicle and charging infrastructure for wireless power transfer, as one system, compatible to 11 kW. The SAE J2954 alignment technology gives additional parking assistance, even allowing vehicles to park and charge themselves autonomously. The SAE task force coordinated with industry and international standards organizations to ensure global WPT harmonization.

Current Status of Charging Stations

Currently, the US Department of Energy has a database of charging station locations in the United States (DOE, 2020b). The DOE website claims that about tens of thousands of EV charging stations are available in the United States. When a user inputs "Electric" (as the fuel type of his vehicle) and his location (city, state, and street address), the website provides a map and list of charging stations with available chargers and connectors.

CONCLUDING REMARKS

This chapter covered many details associated with the problems in recharging EVs and the standardization of infrastructures associated in implementation of future EVs. With improvements in EV technologies and reduction in anticipated prices of EVs, the demand and uses of EVs are more likely to increase rapidly over the next few decades. The increased energy efficiencies and reduction in pollution potential are also major advantages. Further with the implementation of additional features of smart grid technologies will also provide many benefits over the future years. As these new technologies are developing, coordination and integration between different systems and products used in the EV transportation and its infrastructure is needed for increased convenience. The tasks associated in the coordination and successful implementation of the EVs are huge because they are spread between many disciplines (including ergonomics), industry sectors such as transportation, energy generation, energy distribution and modernization of the smart grid. Thus, coordinated efforts between the auto industry, energy companies and government organizations are needed.

REFERENCES

Bablo, J. 2016. *Electric Vehicle Infrastructure Standardization.* EVS29 Symposium, Montreal, Canada. June 19–22, 2016.

Besterfield, D. H., Besterfield-Michna, C., Besterfield, G. H. and M. Besterfield-Scare. 2003. *Total Quality Management.* Third Edition. Upper Saddle River, NJ: Prentice Hall. ISBN 0-13-099306-9.

Bhise, V. D. 2023. *Designing Complex Products with Systems Engineering Processes and Techniques.* Second Edition. Boca Raton, FL: CRC Press.

DOE. 2023. *All Electric Vehicles.* www.fueleconomy.gov/feg/evtech.shtml (accessed August 12, 2023).

DOE. 2020a. *Vehicle Charging.* Energy Efficiency and Renewable Energy Office, Department of Energy. www.energy.gov/eere/electricvehicles/vehicle-charging (accessed: August 20, 2020).

DOE. 2020b. *Electric Vehicle Charging Station Locations.* https://afdc.energy.gov/fuels/electricity_locations.html#/find/nearest?fuel=ELEC (accessed October 13, 2020).

DOE. 2020c. *Developing Infrastructures to Charge Plug-in Electric Vehicles.* https://afdc.energy.gov/fuels/electricity_infrastructure.html (accessed October 13, 2020).

Gold, Russell. 2020. For Electric Cars, California Needs a Bigger Grid. *The Wall Street Journal,* September 26–27, 2020.

International Electrotechnical Commission (IEC). 2017. Electric vehicle conductive charging system – Part 1: General requirements. IEC 61851-1:2017.

NHTSA. 2022. Occupant Protection for Vehicles with Automated Driving Systems. A Rule by the National Highway Traffic Safety Administration published in the Federal Register dated March 30, 2022.

NHTSA. 2023. Federal Motor Vehicle Safety Standards. Federal Register. www.ecfr.gov/current/title-49/subtitle-B/chapter-V/part-571 (accessed: June 10, 2023).

SAE. 2023. *SAE Standards.* SAE International (formerly Society of Automotive Engineers, Inc.), Warrendale, PA.

18 Ergonomic Issues in Autonomous Vehicles

INTRODUCTION

WHAT ARE SELF-DRIVING VEHICLES?

Autonomous vehicles are self-driving (driverless) vehicles that have capabilities to drive without any input or intervention from the drivers. These vehicles have sensing capabilities to continuously monitor the roadway, traffic situations and environmental conditions and make necessary lateral (steering) and longitudinal (accelerator and brake pedal actions) control actions. With integrated GPS support, these vehicles can also select routes and reach preprogrammed destinations. With the implementation of such technologies, the vehicle becomes "autonomous" (i.e., acting separately from other things or people; having the power or right to govern itself).

Currently many companies are offering automotive products with varied levels of automation with the ability for the driver to take over the vehicle controls. Many companies such as Waymo, GM, Ford and others have been testing their research versions of the autonomous vehicles on various public roads to determine if the vehicles can provide safe transportation without any intervention by the test drivers in these vehicles. Some companies have plans to implement the autonomous vehicles on public roads with controlled access (e.g., where the roads have been mapped with more details and/or equipped with additional information transmission capabilities) to enhance their sensing and information acquisition capabilities.

CURRENT STATE

Many of the currently available driver assistance systems such as automatic braking, adaptive cruise control, lane keeping systems are being integrated to create self-driving cars.

The future of such technologies is currently debated because drivers may not be ready to trust such systems and the current versions of the systems are not completely safe. Further, the problem of hacking into such cars needs to be solved to the highest degree of confidence because if the hackers can get into the electronic systems of such vehicles, they can alter output actions of the vehicles. It is expected that in the near future, automakers will integrate many of the driver assistance capabilities and

DOI: 10.1201/9781003485605-5

offer vehicles with limited capabilities (semi-automated and not fully automated self-driving vehicles) such as: a) adaptive cruise control with lane changing capabilities, b) self-parking vehicles, c) auto-pilot features that allow drivers to take their hands off the wheel under certain pre-approved conditions (Naughton, 2015).

Self-driving trucks are another important application area for this technology. Many commercial trucks including those of the army can use self-driving trucks. Sedgwick (2016) describes how the army can benefit from having a convoy of self-driving trucks that can follow a lead truck with a human driver. The potential for reducing driver workload and number of human drivers is also very appealing for many commercial delivery applications. The self-driving trucks can operate over long distances with much less breaks (no coffee breaks and only stop to refuel) and thus can transport cargo in shorter delivery times.

MAJOR SYSTEMS OF AN AUTONOMOUS VEHICLE

The autonomous vehicles have the following three major systems:

1. Global Positioning System (GPS) to locate position of the vehicle on a roadway
2. System to recognize dynamic conditions on the roadway (i.e., cameras and sensors to recognize roadway features, targets, their locations and movements on the roadway)
3. Processors that convert the data from these systems into actions to control lateral and longitudinal movements of the vehicle.

These vehicles include an integrated set of cameras, computers, sensors, mapping software, and radars that work together to take on the sensory and computing abilities possessed by a human driver. The software and hardware components communicate with one another and process millions of points of data each second in order to perform the following operations:

a) Sense the surrounding environment
b) Consider the particular traffic rules and laws in the area
c) Steer, brake, and accelerate the vehicle by use of the control actuators
d) Watch for and react to obstacles
e) Navigate the vehicle to its destination

Multiple Sensors: The sensors used in these vehicles include the Li-DAR (Light Detection and Ranging sensor), which determines the distances to objects (see Figure 18.1). It sends pulsed laser light and measures the time it takes to return. Thus, it measures reflected light back to the sensor. The advantage of lidar is that it can generate precise three-dimensional images of everything from cars to trees to cyclists in a variety of environments and under a variety of lighting conditions. The LiDARs are mounted on the vehicle rooftops and continuously rotate to capture a 360-degree field of view around the vehicle with spinning mirrors and electric motors. The mechanical

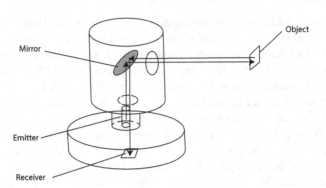

FIGURE 18.1 LiDar measures the time between the emitted pulsed laser signals and the sensed reflected signals from objects.

lidar will be replaced in the future with three 120-degree solid state lidars. The solid-state lidar can direct a signal over a hundred meters of distance with an accuracy of a fraction of a degree. In addition, video cameras in front and rear of the vehicle, radar systems, ultrasonic sensor on the sides to detect presence of objects and vehicles in the adjacent lanes are used to capture data.

Vehicle Communications: A number of vehicle communication channels (e.g., 5G) are now available to facilitate high speed two-way data communications from vehicles to other locations, for example, other surrounding vehicles, cloud/database, infrastructure, home, other persons outside the vehicle (pedestrians) and utility companies (e.g., EV charging) and vehicle manufacturers' data-processing centers. These data provide up-to-date information on situations on the roadways, for example, maps and navigation information, status of their vehicle, other vehicles, traffic, changes in traffic (due to accidents, detours, work zones) and so forth.

According to recent studies, nearly 25 percent of all miles driven in the United States could be shared by autonomous electrical vehicles (SAEVs) by 2030. Electric powertrain is indispensable for autonomous vehicles as it offers a) higher fuel efficiency and reduced CO_2 emissions, b) an easier platform to support drive-by-wire systems needed for vehicle autonomy, and c) an attractive proposition of lower cost of ownership and maintenance, especially for fleet owners in ride-sharing systems. However, the integration of vehicle autonomy with electrification is not going to be a simple additive manufacturing task. For example, autonomous functionality-enabling electronics power in a fully automated autonomous car can demand up to 2–4 kW. This power demand when fed by the main battery can reduce power available to drive the vehicle significantly (up to 35%, especially in city driving). However, other factors such as smoother driving profile of a connected autonomous vehicle compared to human driving, less time spent in vehicle refueling and maintenance can help enhance electric vehicle drive range. Automakers and suppliers need to account for these tradeoffs from early concept design phase for an autonomous electric vehicle (Siemens, 2022).

The automakers have a lot to consider while designing future vehicles. The trend towards software-defined vehicles is enabling new and important breakthroughs for autonomous, connected, electric, and shared (ACES) mobility. As these vehicles get increasingly smarter and more automated, the vehicle software platforms supporting the increasing number of software functions are becoming far more complex and sophisticated. The challenge for automotive manufacturers is to support autonomous innovation in every way possible, and to speed up time to market for new features. The interdependencies between safe driving features and entertainment features will require a holistic approach to vehicle systems management. The potential for offering driver "delights" (i.e., "Wow" features in the Kano model of quality) to differentiate vehicle brands can be increased. However safety must remain a non-negotiable vehicle attribute to meet customer needs and trust. The increased connectivity also exposes vehicles to many external factors that increase many new risks to the auto manufacturers – for example, system failures due to missed or errors in detections can have large consequences such as loss of customer trust and increased product liability claims.

Auto manufacturers have opportunities to focus on evolving needs for onboard information and entertainment systems. However, delivering the information and entertainment features and customer wants requires efficient management of many considerations involving multiple disciplines. At the same time, the manufacturers are under pressure to bring new information and entertainment capabilities to market very quickly, which means that the entire product development process must be agile and efficient.

As systems become more complex, the safety challenges also increase. For example, aspects that directly affect safety, such as vulnerability to specific environmental factors, must be considered and suitable solutions implemented, and cybersecurity challenges associated with connected vehicles must be adequately addressed. Trust is key and the customers expect that the in-vehicle devices are safe and reliable under all operating situations.

SAE LEVELS OF AUTOMATION

SAE International, formerly the Society of Automotive Engineers, have published a standard called "Levels of Automation (J3016)" (SAE, 2014), which defines the following six levels of automation that can be used in automotive products.

Level 0: No Automation: This condition involves full-time driver involvement with all aspects of the dynamic driving task, even when the driver is provided with warning or momentary assistance or intervention systems (e.g., automatic emergency braking system, blind area warning system).

Level 1: Driver Assistance: In this driving mode specific execution of a driver assistance systems of steering (e.g., lane centering control system) or longitudinal control (acceleration/deceleration) using information about the driving environment (e.g., an adaptive cruise control which keeps track of other vehicles in the front) with the expectation that the driver will perform all remaining aspects of the dynamic driving task. However, the driver is required to intervene and take over the functions of the assistance systems anytime.

Level 2: *Partial Automation:* In this driving mode specific execution of one or more driver assistance systems of both steering (e.g., lane centering control system) and acceleration or deceleration using information about the driving environment (e.g., an adaptive cruise control) with the expectation that the driver will perform all remaining aspects of the dynamic driving task. However, the driver is required to intervene and take over the functions of the assistance systems at any time.

Level 3: *Conditional Automation:* In this driving mode specific execution of an automated driver assistance system of all aspects of the dynamic driving task. When the driver assistance feature requests, the driver must respond appropriately and take over control of the vehicle. An example of such systems is a traffic jam chauffeur.

Level 4: *High Automation:* In this driving mode specific performance by an automated driving system of all aspects of the dynamic driving task even if the human driver does not respond appropriately to a request to intervene. For example, a local driverless taxi can only function automatically without any human driver involvement when certain specific roadway and driving conditions are met.

Level 5: *Full Automation:* In this mode the full-time performance of an automated driving system of all aspects of the dynamic driving task under all roadway and environmental conditions that can be managed by a human driver. This condition is the same as level 4, but all driving tasks can be handled everywhere under all conditions by the automated system. For example, a robot taxi on any roadway and in any environmental condition. No human driver is needed.

Currently most driverless vehicles are capable of achieving levels 2 or 3 automation, which will require a human driver who can intervene when the automated system cannot control the vehicle. A few driverless vehicles can operate at level 4 automation when additional information transmission devices are mounted along the roadway to provide route guidance and control information.

There is not a fully autonomous (Level 5) vehicle for sale in the market today, but some automakers are advancing in the field. Over the past few years, a few automotive products have been introduced in the market with driver assistance features that take much of the burden off the driver. These systems alleviate driver fatigue by assisting with steering and acceleration. This technology is beneficial for drivers with long commutes. And it can come in handy during road trips requiring lots of driving.

BENEFITS OF DRIVERLESS VEHICLES

1. No driver will be required. Thus, driver training and licensing may not be needed. Individuals who do not like or want to drive will not be required to drive. Drivers who like to drive may select a vehicle that can be driven in different levels (fully or partially) of driver-controlled or driverless modes.

2. According to a 2013 report by McKinsey & Company, the benefits of driverless cars – including improved safety, time savings, productivity increases,

and lower fuel consumption and emissions – could have a total economic impact of $200 billion to $1.9 trillion per year by 2025.

3. Automobile manufacturers will benefit, as will the dozens of companies supplying microprocessors, cameras, sensors, and mapping software that allow driverless cars to operate autonomously.

4. The average citizen stands to benefit, too, in the form of increased productivity, fuel savings, and lower insurance premiums.

5. The efficiency of autonomous vehicles coupled with rapid advances in electric and alternative fuels will generate significant environmental benefits.

DISADVANTAGES OF DRIVERLESS VEHICLES

1. Job losses for driver professions (e.g., delivery/truck drivers)
2. Reduction in vehicle sales, oil demand, parking spaces, and insurance costs
3. Reduced number of accidents, fatalities and organ donations
4. Reduced number of traffic tickets and revenues from tickets and fines
5. Reduced driving opportunities for drivers who enjoy driving or need a vehicle to travel immediately without any waiting time

NEAR FUTURE OF DRIVERLESS VEHICLES

1. It is expected that in the near future, the automakers will integrate many of the driver-assistance capabilities and offer vehicles with limited capabilities (semi-automated and not fully automated self-driving vehicles) such as: a) adaptive cruise control with lane changing capabilities, b) self-parking vehicles, c) auto-pilot features that allow drivers to take their hands off the wheel under certain pre-approved conditions (Naughton, 2015).

2. On March 10, 2022, NHTSA published a final rule amends the occupant protection federal motor vehicle safety standards (FMVSSs) to account for future vehicles that do not have the traditional manual controls associated with a human driver because they are equipped with Automated Driving Systems (ADS). The final rule makes clear that, despite their innovative designs, vehicles with ADS technology must continue to provide the same high levels of occupant protection that current passenger vehicles provide.

ERGONOMIC CONSIDERATIONS IN DESIGNING AUTONOMOUS VEHICLES

The driver is expected to be ready and available to intervene with the automated controller until the automation 4 and 5 level systems are perfected, and drivers trust the operability of the system. Until then the following ergonomic issues need considerable research in information processing and decision making:

1. *Vigilance capabilities*: Drivers have limited capabilities in monitoring the autopiloting performance over long periods of time. Literature on vigilance

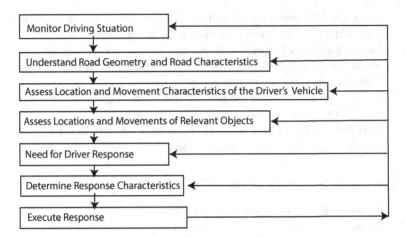

FIGURE 18.2   Flow diagram of driver's information processing activities.

(i.e., long-term ability of human operators to monitor equipment in detecting signals over long term periods of time) has shown that humans are not very good at such tasks and their performance degrades after about half-hour of continuous watch. (Mackworth, 1961; Wickens, Gordon and Liu, 1998,)

2. *Understanding and programming driver information processing*: During normal driving (without any driver assistance features in the vehicle), the driver continuously samples visual information by making eye movements and obtains information on many successive tasks (shown in Figure 18.2). Extensive future research in the driver information processing abilities is needed to determine how micro-processors in automated vehicles should be designed and programmed to create even safer automated vehicles than those currently operated by the drivers.

3. *Updating databases*: Databases on changes in roadways, diversions (detours), maps, traffic movements need to be continuously updated on a real-time basis to avoid misdirection of vehicle control actions. Temporary changes in work zones must also notify the vehicle and the vehicle should be capable of detecting work zones and responding to changes in traffic flows.

4. *Changing weather and lighting conditions*: The vehicle sensors and sensed data processors must have the capability to accurately identify driving situations under all possible weather and lighting conditions. This means that future highway infrastructure must transmit roadway and vehicle location information to the vehicle in adverse weather conditions.

5. *Predicting actions of drivers in other vehicles in close vicinity*: The automated vehicle must be able to sense all surrounding vehicles and accurately predict their possible maneuvers.

6. *Predicting and executing vehicle response*: The automated vehicle should be able to monitor and respond to maneuvers of other vehicles in the close vicinity and maneuver the automated vehicle safely (meeting all traffic rules) throughout its trip.

Ergonomic Considerations to Meet Other Customer Needs

1) *Easy to enter and egress*: Automated vehicles should provide steps and/or running boards (if floor height is too high) and grab handles or hand bars.

2) *Safe*: Automated vehicles should not cause any injuries to occupants during all uses. The automated vehicles must meet NHTSA's all applicable safety standards (i.e., must meet crash avoidance requirements, e.g., vehicle lighting braking and vehicle stability and impact protection requirements, e.g., provide seat belts and air bags (NHTSA, 2022).

3) *Accommodate occupants*: The automated vehicles must provide the required space for occupants in terms of headroom, legroom, shoulder room, and so forth.

4) *Provide comfortable seats*: The automated vehicle should accommodate a large percentage of passengers with adjustable recline angle, seat height, armrests, and lumbar support, reduce road induced vibrations and provide thermal comfort (e.g., heated and cooled seats).

5) *Provide visibility*: Passengers in the automated vehicles should be able to see outside fields to the front, rear and sides of the vehicle for enjoyment and a safe and comfortable feeling.

6) *Provide Wi-Fi access for communications*: The automated vehicles should provide wireless communication to their occupants.

7) *Provide automatic climate control*: Automated vehicles should provide thermal comfort for their occupants.

8) *Provide emergency control*: Automated vehicles should allow capability to abort the ride (i.e., ability to safety pull on the side of the road when possible or choose a different destination).

9) *Provide legible screens and easy to obtain displayed information*: Automated vehicles should display information such as speed, current location, next stop, landmarks or points of interests along the route (see Volume 1, Chapter 7) to its occupants from their seated positions.

10) *Provide auditory information on vehicle destination, route and points of interest*: Automated vehicles should present voice messages to the passengers related to the trip at right time for the passengers to confirm that the right vehicle has arrived and to get ready for their entry or exit from the vehicle.

CONCLUDING REMARKS

Designing driverless vehicles is a major challenge and a technical breakthrough. It is easier to produce a vehicle with the SAE automation levels below 4 when a driver can take over controls of the vehicles when the automated features are unable to respond safely. The vehicles designed for automation levels below 4 will need to accommodate a human driver and provide all interfaces (i.e., controls, displays and body support). However, the automated vehicles that can operate at automation levels 4 and 5 do not need to accommodate a human driver and no human intervention is required from a human inside the vehicle. Thus, such automated vehicles would not provide interfaces for the driver. The fully automated vehicles need to be monitored and controlled continuously (by computers within and outside the vehicle) to ensure that they do not perform any unsafe maneuver. The coordinated control of mix of many automated vehicles on the highway along with other vehicles with different levels of automation (below 4) would also be a major challenge.

REFERENCES

Mackworth, N. H. 1961. Researches on the Measurement of Human Performance. In Selected *Papers on Human Factors in the Design and Use of Control Systems*, ed. H. Wallace Sinaiko, 174–331, London, UK: Dover Publications, Inc.

NHTSA. 2022. Occupant Protection for Vehicles with Automated Driving Systems. A Rule by the National Highway Traffic Safety Administration published in the Federal Register dated March 30, 2022.

Siemens. 2022. White Paper: Design Challenges and Opportunities for Electric Powertrain with Vehicle Autonomy. https://bit.ly/3PJksF2 (accessed: July 25, 2022).

TTTECH Auto. 2022. https://bit.ly/3TFiQ0l (accessed: March 17, 2024)

Wickens, C. D., S. E. Gordon and Y. Liu. 1998. *An Introduction to Human Factors Engineering*. ISBN: 0.321-01229-1. New York, NY: Addison-Wesley Educational Publishers, Inc.

19 Vehicle Evaluation Methods

OVERVIEW ON EVALUATION ISSUES

An automotive product is used by a number of users in a number of different usages. To ensure that the vehicle being designed will meet the needs of its customers, the ergonomic engineers must conduct evaluations of all vehicle features that involve human interfaces under all possible usage situations. Usage can be defined in terms of each task that needs to be performed by a user to meet a certain objective. A task may have many steps or subtasks. For example, the task of getting into a vehicle would involve a user performing a series of subtasks such as: a) unlocking the door, b) opening the door, c) entering the vehicle and sitting in the driver's seat, and d) closing the door.

The ergonomic evaluations are conducted for a number of purposes, such as a) to determine if the users will be able to use the vehicle and its features, b) to determine if the vehicle has any unacceptable features that will generate customer complaints after its introduction, c) to compare the user preferences for a vehicle or its features with similar features in other vehicles, and d) to determine if the product will be perceived by the users to be the best in the industry (i.e., that it will perform better than other vehicles in the same class or market segment).

The purpose of this chapter is to review methods that are useful in ergonomic evaluations of vehicles. The evaluations can be conducted by collecting data in a number of situations. Examples of data collection situations are given below:

1. A product – vehicle or one or more of its systems, chunks (portion of the vehicle, e.g., front end, rear end) or features – is shown to a user (an evaluator) and the user's responses (e.g., facial expressions, verbal comments) are noted (or recorded). This situation occurs when a concept vehicle is displayed in an auto show.
2. A product is shown to a user, and then responses to questions asked by an interviewer are recorded. This situation occurs in a market research clinic.
3. A customer is asked to use a product, and then, responds to a number of questions asked by an interviewer (usually by using a predeveloped questionnaire), which are recorded. This situation can occur in a drive evaluation.

DOI: 10.1201/9781003485605-6

4. A user is asked to use a number of products, and the user's performance in completing a number of tasks on each of the products is measured. This situation can occur in a performance measurement study using a set of vehicles – or alternate designs of a vehicle system – in test drives.

5. A user is asked to use several products and then asked to rate the products based on a number of criteria (e.g., preference, usability, accommodation, and effort). This situation occurs in field evaluations using the manufacturer's test vehicle and other competitive vehicles.

6. A sample of drivers are provided with instrumented vehicles that record vehicle outputs and video data of driver behavior and performance as the participants drive where they wish, as they wish, for weeks or months each. This is probably the only valid method to discover what drivers actually do over time in the real world. (This situation occurs in naturalistic driving behavior measurement studies.)

The above examples illustrate that an ergonomics engineer can evaluate a vehicle or its features by using a number of data collection methods and measurements.

ERGONOMIC EVALUATIONS DURING VEHICLE DEVELOPMENT

During the entire vehicle development process, many evaluations are conducted to ensure that the vehicle being designed will meet the needs of the customers. The design issues and ergonomic considerations covered in all the chapters in this book need to be systematically evaluated to ensure that all design requirements are considered, and that appropriate evaluation methods are used. The results of the evaluations are generally reviewed in the vehicle development process at different milestones with various design and management teams.

Table 19.1 provides a summary of ergonomic evaluations and the evaluation methods used in the entire vehicle development process. The systems engineering model provided in Volume 1, Figure 2.5 is used here to provide reference to the timings of different events in the vehicle development process. The left two columns in the table present the order and brief description of the general areas that need evaluation. The middle columns provide types of evaluation methods used during the vehicle development process. The timings of the evaluations are indicated by two or three letter codes that refer to the part of the systems engineering "V" model presented in the lower portion of the table. The right-hand column presents some details of the evaluation methods, requirements and issues to be addressed in each row. The second to last column provides chapter number where these areas are covered in this book.

EVALUATION METHODS

Table 19.2 provides a summary of methods categorized by combinations of types of data-collection methods and types of measurements.

The left-hand column of Table 19.2 shows that the data can be collected by using methods of observation, communication and experimentation. In the observation

TABLE 19.1
Summary of Ergonomic Evaluations and the Evaluation Methods Used in the Vehicle Development Process

No.	Vehicle Evaluation Need	Type of Evaluation Methods and Their Application Timings* in the Vehicle Design Process					Chapter Number to Refer for Additional Info.	Some Details on Evaluation Methods and Issues
		Checklists and Judgments of Experts	Applications of Data-based Requirements, Models and Standards	Static Tests -- Laboratory and Simulators	Drive Tests and Evaluations			
1	Vehicle Specification and Attribute Requirements Development	SRT	LE				2, 3, 13, 19	Benchmarking competitors vehicles, conduct market research clinic to understand customer needs. Conduct QFD studies. Develop Package and Ergonomics attribute requirements.
2	Driver Positioning, Primary Controls and Occupant Locations	SRT	LE				5,6,7,19	CAD/CAE Applications of SAE standards J1516, J1517, J4002, J4003, J4004, J287, J1050, J1052; Special Populations

(continued)

TABLE 19.1 (Continued)
Summary of Ergonomic Evaluations and the Evaluation Methods Used in the Vehicle Development Process

No.	Vehicle Evaluation Need	Type of Evaluation Methods and Their Application Timings* in the Vehicle Design Process				Chapter Number to Refer for Additional Info.	Some Details on Evaluation Methods and Issues
		Checklists and Judgments of Experts	Applications of Data-based Requirements, Models and Standards	Static Tests -- Laboratory and Simulators	Drive Tests and Evaluations		
3	Available Space to Locate Controls and Displays	LL	LE			7	Controls and Display Concepts and Location Considerations; CAD Applications of SAE J941, J287, Minimum Reach, Down Angle Requirement, and J1050
4	Entry/Exit Evaluations for Body Door Openings, Seats, Steering Wheel and Armrest	LE		LM		10	Historic Data from Customer Responses; CAD Applications of Digital Manikins; Task Analyses; Subjective Evaluation Using Rating Scales

5	Field of View Evaluations	LM	LM		8	Historic Customer Response; CAD/ CAE Evaluations of FMVSS 103, 104, 111, SAE J1050
6	Instrument Panel, Door Trim Panel and Console	LM	LM		7, 12, 16	Location Requirements in CAD, SAE J1138, J1139, FMVSS 101
7	Interior Buck Evaluation	LE	LM		5, 19	SAE J826; CAD/ CAE Manikin Based Evaluations; Subjective Ratings
8	Sunlight Reflections, Veiling Glare, Legibility Evaluations	LM	LM	RM	8, 14	CAD Ray Tracing Analyses of Reflections; Veiling Glare Analysis; Subjective Assessments of Sunlight Reflections (outdoors and in daylight simulation rooms)
9	Controls and Displays Operability	LE	LM	RM	4, 5, 7, 14, 15, 16	Driver Workload Evaluations in Simulators and Field Tests with Prototypes; Human Error Analyses; FMEA

(continued)

TABLE 19.1 (Continued)
Summary of Ergonomic Evaluations and the Evaluation Methods Used in the Vehicle Development Process

| No. | Vehicle Evaluation Need | Type of Evaluation Methods and Their Application Timings* in the Vehicle Design Process | | | | Chapter Number to Refer for Additional Info. | Some Details on Evaluation Methods and Issues |
		Checklists and Judgments of Experts	Applications of Data-based Requirements, Models and Standards	Static Tests -- Laboratory and Simulators	Drive Tests and Evaluations		
10	Interior Lighting and Graphics Evaluations		LM	RM	RL	14	Visibility and Legibility Models; Subjective Assessments of Interior Lighting and Lighted Components for Appearance, Harmony and Legibility.
11	Vehicle Lighting	LE	LL	RM	RM	9, 14	Technology Feasibility; Visibility Prediction Models; FMVSS108 and SAE Lighting standards; Night Drives to Evaluate Pleasing Perceptions of Beam Patterns; Subjective Assessments of Lighted Appearance of Signal Lamps

#	Method					#	Description
12	Engine Service Evaluations	LE	LM		RM	11	Task Analyses; Evaluation of Labels and Hand Clearances
13	Trunk Space and Cargo Loading/ Unloading	LE	SRT		RM	11	SAE J1100 Luggage/ Cargo volumes; Biomechanical Models and Task Analyses
14	Craftsmanship	LL		RL	RL	12	Evaluation of Pleasing Perceptions --Fits/ Gaps, Materials, Textures and Color Harmony, etc.
15	Final Drive Evaluations (Entire Vehicle)				RL	19	Checklists, Subjective Evaluations by Experts, Customers and Management Personnel

Note: * = timing within the Systems Engineering "V" Model (see Figure on the right)

SRT = Start of the program
L = Left (throughout left side)
LE = Left side (of "V") early
LM = Left side middle
LL =Left side late
R = Right (throughout right side)
RE =Right side early
RM = Right side middle
RL = Right side late
END = End of the program

TABLE 19.2
Types of Data Collection and Measurement Methods

Type of Data Collection Method	Type of Measurements	
	Objective Measurements	Subjective Measurements
Observation	Experimenter Observed or Data Recorded with Instruments; Behavior Observations (Glances, Durations, Errors, Difficulties, Conflicts), Near-accidents	Checklists Completed by Subjects Based on Their Observations
Communication	Experimenter Reported Objective Measures (e.g. speed, events)	Subject Reported --Detections, Identifications; Responses in Checklists; Responses (or ratings) Using Nominal, Interval and Ratio Scales; Problems, Difficulties and Errors during Operation of Equipment
Experimentation	Measurements with Instrumentation: Performance Measurements, Behavioral Measurements	Obtained from Subjects: Ratings, Behavioral Measurements (Difficulties, Errors, etc.)

method, it is assumed that a subject performing a task can only be observed by an experimenter or the data can be recorded (e.g., by using a camera) for later observations by an analyst or an experimenter. In the communication method, the subject (or the experimenter) can be asked to report about the problems experienced while performing a task or asked to provide ratings on his/her impressions about the task. In the experimentation method, the test situations are designed by deliberate changes in combinations of certain independent variables (e.g., configuration of the product) and the responses are obtained by using combinations of methods of observation and/or communication. For further information on many available methods of data collection and their advantages and disadvantages, the reader should refer to Chapanis (1959) and Zikmund and Babin (2009).

The types of measurements can be categorized as objective or subjective as shown in Table 19.2. The objective measures can be defined here as measurements that are not affected by the subject performing the tasks or by the experimenter observing or recording the subject's performance. The objective measures are generally obtained by use of physical instruments or by unbiased and trained experimenters. Subjective measurements are generally based on the subject's perception and experience during or after performing one or more tasks. The objective measures are generally preferred because they are more precise and unbiased. However, there are many vehicle attributes that cannot be measured without using human subjects as "measuring instruments". After the users have experienced the vehicle, they are better able to express their perceived impressions about the vehicle and its characteristics by the use of methods of communication.

The following section provides additional information on the methods of data collection relevant to ergonomic evaluations.

METHODS OF DATA COLLECTION AND ANALYSIS

OBSERVATIONAL METHODS

In observational methods, information is gathered by direct or indirect observations of subjects during their product usages to determine if the product is easy or difficult to use. An observer can directly observe, or a video camera can be set up and its recordings can be played back at a later time. The observer needs to be trained to identify and classify different types of predetermined behaviors, events, problems or errors that a subject commits during the observation period. The observer can also record durations of different types of events, number of attempts made to perform an operation, number and sequence of controls used, number of glances made, and so forth. Some events such as accidents are rare, and they cannot be measured through direct observations due to excessive amount of direct observation time needed until sufficient accident data are collected. However, information about such events can be obtained through reports of near-accidents (i.e., situations where accidents almost occurred but were averted) and indirect observations (e.g., through witnesses, or from material evidence) gathered after such events. Therefore, the information gathered through indirect observations may not be very reliable due to a number of reasons (e.g., witness may be guessing or even deliberately falsifying; or objects associated with the event of interest may have maybe been displaced or removed).

COMMUNICATION METHODS

The communication methods involve asking the user or the customer to provide information about his or her impressions and experiences with the product. The most common technique involves a personal interview where an interviewer asks the user a series of questions. The questions can be asked prior to usage of the product, during usage or after usage. The user can be asked questions that will require the user to: a) describe the product or the impressions about the product and its attributes (e.g., usability); b) describe the problems experienced while using the product (e.g., difficult to read a label); c) categorize the product using a nominal scale (e.g., acceptable or unacceptable; comfortable or uncomfortable; liked or disliked); d) rate the product on one or more scales describing magnitude of its characteristics and/or overall impressions (e.g., workload ratings, comfort ratings, difficulty ratings); or e) compare the products presented in pairs based on a given attribute (e.g., ease of use, comfort, quality feel during operation of a control).

Commonly used communication methods in product evaluations include: 1) rating scales: using numeric scales, scales with adjectives (e.g., acceptance ratings and semantic differential scales); 2) paired comparison based scales (e.g., using

Thurstone's Method of Paired Comparisons and Analytical Hierarchical method) described later in this chapter.

In addition, many tools used in fields such as Industrial Engineering, Quality Engineering and Design for Six Sigma, and Safety Engineering can be used. Some examples of such tools are process charts, task analysis, arrow diagrams, interface diagrams, matrix diagrams, quality function deployment (QFD), Pugh analysis, failure modes and effects analysis (FMEA), and fault tree analysis (FTA). The above mentioned tools rely heavily on the information obtained through the methods of communication from the users/customers and members of the multi-functional design teams. Additional information on many of these tools can be obtained from Bhise (2023), Besterfield, Besterfield-Michna, et al. (2003), Creveling and Slutsky (2003) and Yang and El-Haik (2003).

EXPERIMENTAL METHODS

The purpose of experimental research is to allow the investigator to control the research situation (e.g., selecting a vehicle design, test condition) so that causal relationships between the response variable and independent variables that define the vehicle characteristics (e.g., interface configuration, type of control, type of display, operating forces) may be evaluated. An experiment includes a series of controlled observations (or measurements of response variables) undertaken in artificial (test) situations with deliberate manipulations of combinations of independent variables in order to answer one or more hypotheses related to the effect of (or differences due to) the independent variables. Thus, in an experiment one or more variables (called independent variables) are manipulated and their effect on another variable (called dependent or response variable) is measured, while all other variables that may confound the relationship(s) are eliminated or controlled.

The importance of the experimental methods is that a) they help identify the best combination of independent variables and their levels to be used in designing the vehicle and thus provide the most desired effect on the users, and b) when the competitors' products are included in the experiment along with the manufacturer's product, the superior product can be determined. To ensure that this method provides valid information, the researcher designing the experiment needs to ensure that the experimental situation is not missing any critical factor related to performance of the product or the task being studied. Additional information on the experimental methods can be obtained from Kolarik (1995) or other textbooks on Design of Experiments.

Examples of Some Experimental Techniques

   a) *Observational Studies*: Driver and customer observational studies are conducted to obtain information on issues such as problems encountered while entering and exiting vehicles (see Volume 1, Chapter 10 and Bodenmiller et al., 2002.), operating in-vehicle devices (e.g., to study driver understanding of various control functions in audio, climate controls and navigation systems), and performing vehicle service tasks (e.g., checking fluids, changing fuses and bulbs, refueling, changing a tire).

b) *Studies Using Programmable Vehicle Bucks*: Programmable vehicle bucks are used in early package evaluation studies to assess exterior and interior dimensions such as vehicle width, windshield rake angle, seating reference point location, driver eye location, visibility over the instrument panel, hood and side windows, height of armrest, and so forth (Richards and Bhise, 2004).

c) *Driving Simulator Studies*: Driving simulators are now routinely used in many automotive companies to evaluate driver workload issues in operating various in-vehicle devices (Bertollini et al., 2010). All three methods of observation, communication and experimentation can be used during the simulator tests.

d) *Field Studies and Drive Tests*: Various studies under actual driving situations on test tracks and public roads under different road, traffic, lighting and weather conditions are conducted for evaluation of issues in areas such as seat comfort, field of view, vehicle lighting, controls and displays usage, and driver workload (Jack et al., 1995, Owens et al., 2010; Tijerina et al., 1999).

OBJECTIVE MEASURES AND DATA ANALYSIS METHODS

Depending upon the task used to evaluate a product, task performance measurement capabilities and instrumentation available, the ergonomics engineer would design an experiment and procedure to measure dependent measures. The objective measures can be based on temporal, physical or spatial measures such as time (taken or elapsed), distance (position or movements in lateral, longitudinal or vertical directions), velocities, accelerations, events (occurrences of predefined events), and measures of user's physiological state (e.g., heart rate). The recorded data are reduced to obtain the values of the dependent measures and their statistics such as means, standard deviations, minimum, maximum, percentages above and/or below certain preselected levels. The measured values of the dependent measures are then used for statistical analyses based on the experiment design selected for the study. Some examples of applications involving objective measures are provided in a later section of this chapter.

SUBJECTIVE METHODS AND DATA ANALYSIS

Subjective methods are used by the ergonomics engineers because in many situations a) the subjects are better able to perceive characteristics and issues with the product, and thus, they can be used as the measurement instruments, b) suitable objective measures do not exist, and c) the subjective measures are easier to obtain.

Pew (1993) has pointed out several important points regarding subjective methods. Subjective data must come from the actual user rather than the designer; the user must have an opportunity to experience the conditions to be evaluated before providing opinions; and care must be taken to collect the subjective data independently for each subject; and the final test and evaluation of a system should not be based solely on subjective data.

The following methods are covered in Volume 1, Chapter 3: a) Benchmarking and Breakthrough, b) Pugh Diagram, c) Quality Function Deployment and d) Failure Modes and Effects Analyses. These methods help evaluate and determine vehicle characteristics during the early phases of the product development process.

The following part of this chapter cover techniques that involve checklists, user interviews involving rating on a scale, paired comparison techniques that develop a rating scale and scale values of the evaluated products or their characteristics

CHECKLISTS

A checklist is used to check if the product being designed meets each applicable ergonomic guideline (or principle or requirement) in the area covered by the checklist. The checklist approach is commonly used during design of many areas, such as: a) interior and exterior package design (refer to requirements in Volume 1, Chapter 5); b) controls and displays design (refer to requirements in Volume 1, Chapter 7); c) vehicle lighting design (refer to requirements in Volume 1, Chapter 9); and d) special population issues (refer to Chapter 24).

The checklists must be comprehensive and complete and must be completed by trained evaluators. The ergonomic checklists are generally completed by ergonomics experts based on their knowledge or data available from various ergonomic analyses and studies (see Chapters 15 and 16). Pew (1993) has compiled a useful checklist of "poor questions" that should guide the development of any checklist or questionnaire. Some examples of poor questions show they a) produce a narrow range of answers, b) they require information the respondent does not know or remember, and c) their statement is too vague.

VEHICLE USER INTERVIEWS

Drivers and other vehicle users are interviewed individually and in groups (e.g., focus group sessions) to understand their concerns, issues and wants related to various vehicle features. For example, Bhise, Hammoudeh, Dowd and Hayes (2005) asked drivers to develop layouts of center stack and console areas through a structured interview technique (a method of communication).

The two most commonly used subjective measurement methods used during the vehicle development process are a) rating on a scale and b) paired comparison based methods. These two methods are presented below.

RATING ON A SCALE

In this method of rating, the subject is first given instructions on the procedure involved in evaluating a given product including explanations on one or more of the product attributes and the rating scales to be used for scaling each attribute. Interval scales are used most commonly. Many different variations are possible in defining the rating scales. The interval scales can differ due to; a) how the end points of the scales are defined; b) number of intervals used (Note: odd number of intervals allow use of a mid-point); and c) how the scale points are specified (e.g., without descriptors versus with word descriptors or numerals).

Figure 19.1 presents eight examples of interval scales. The first four scales ('a' through "d") are numeric scales with end points defined by descriptors (words or adjectives). The first two scales have 10 points, and their numeric values range from

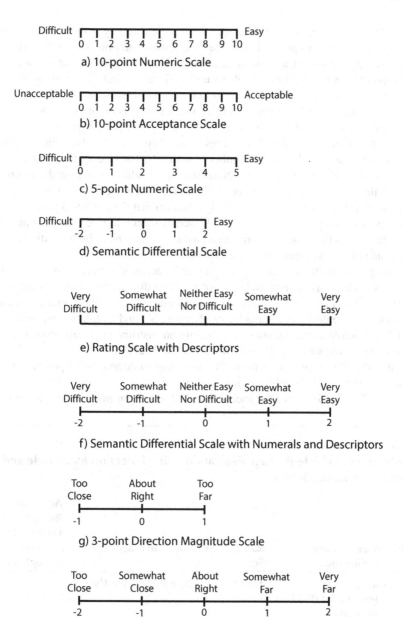

FIGURE 19.1 Examples of rating scales.

0 to 10. Whereas the remaining scales ("d" through "h") have clearly defined mid-points and numbers; and/or adjectives (or descriptors) are used in defining each scale marking. The use of adjectives or descriptors can help subjects in understanding the levels of the attribute associated with the scale. The use of mid-points (e.g., in scales

"e" or "f") allows the subject to choose the middle category if the subject is unable to decide whether the product attribute in question falls on one or the other side of the scale. The use of a scale such as "e" also allows the subject to first decide if the product was easy or difficult to use, and then select the level by using the adjectives "somewhat" or "very". Even number of intervals can also be used where the subject would be forced to decide between either sides of the scale. Thus, the mid-point associated with the inability of the subject to decide between the two sides will be removed. The scales with 5 or less points are easier for the subjects to use as compared to scales with a larger number of intervals. The direction magnitude scales such as scales "g" and "h" are particularly useful in evaluating vehicle dimensions. In these scales, the mid-point is defined by the words "about right" and thus, a large percentage of responses in this category helps confirm that the evaluated product dimension was designed properly. Whereas a skewed distribution of responses to the left or the right side on the scale will indicate a mismatch in terms of both the direction and magnitude of the problem with the dimension.

Rating methods using different interval scales are used for ergonomic evaluations of issues such as: a) interior and exterior package dimensions; b) characteristics of controls and displays (e.g., acceptability of locations, sizes, grasp areas, feel during movement of controls, compressibility of armrests); and c) interior materials (e.g., visual and tactile characteristics of materials on instrument panels, door trim, seat areas and steering wheels) (Bhise et al., 2006, 2008, 2009).

Table 19.3 illustrates how 3-point direction magnitude and the 10-point acceptance scales together can be used to evaluate a number of interior dimensions in a vehicle package. The distribution of responses on each direction magnitude scale provides

TABLE 19.3
Illustration of Vehicle Package Evaluation Using Direction Magnitude and Acceptance Rating Scales

Item No.	Driver Package Consideration	Rating Using Direction Magnitude Scale			Acceptance Rating: 1= Very Unacceptable, 10=Very Acceptable
1	Steering wheel longitudinal (fore/aft) location	Too Close	About Right	Too Far	
2	Steering wheel vertical (up/down) location	Too Low	About Right	Too High	
3	Steering wheel diameter	Too Small	About Right	Too Large	
4	Gas pedal fore/aft location	Too Close	About Right	Too Far	
5	Gas pedal lateral location	Too Much to Left	About Right	Too Much to Right	

TABLE 19.3 (Continued)
Illustration of Vehicle Package Evaluation Using Direction Magnitude and Acceptance Rating Scales

Item No.	Driver Package Consideration	Rating Using Direction Magnitude Scale			Acceptance Rating: 1= Very Unacceptable, 10=Very Acceptable
6	Lateral distance between the gas pedal and the brake pedal	Too Small	About Right	Too Large	
7	Gas pedal to brake pedal lift-off	Too Small	About Right	Too Large	
8	Gearshift lateral location	Too Much to Left	About Right	Too Much to Right	
9	Gearshift location longitudinal location	Too Close	About Right	Too Far	
10	Height of the top portion of the instrument panel directly in front of the driver	Too Low	About Right	Too High	
11	Height of the armrest on driver's door	Too Low	About Right	Too High	
12	Belt height (lower edge of the driver's side window)	Too Low	About Right	Too High	
13	Space above the driver's head	Too Little	About Right	Too Generous	
14	Space to the left of the driver's head	Too Little	About Right	Too Generous	
15	Knee space (between instrument panel and right knee with foot on the gas pedal)	Too Little	About Right	Too Generous	
16	Thigh space (between the bottom of the steering wheel and the closest lower surface of the driver's thighs).	Too Little	About Right	Too Generous	

feedback to the designer on how the dimension corresponding to the scale was perceived in terms of its magnitude; and the ratings on the acceptance scale provide the level of acceptability of the dimension. For example, if the ratings on item number 5 (Gas pedal lateral location) in Table 19.3 showed that 80 percent of the subjects rated the gas pedal location as "Too much to the left" on the direction magnitude scale and the average rating on the 10-point acceptance scale was 4.0, the designer can conclude that the gas pedal needs to be moved to the right to improve its accept-ability. The author found that such use of dual scales was very helpful in fine-tuning the vehicle dimensions in the early stages of the vehicle design process.

PAIRED COMPARISON-BASED METHODS

The method of paired comparison involves evaluating products presented in pairs. In this evaluation method, each subject is essentially asked to compare two products in each pair using a predefined procedure and is asked to simply identify the better product in the pair on the basis of a given attribute (e.g., comfort, usability). If the respondent says there is no difference between the two products, the instruction would be to randomly pick one of the products in the pair. The idea is that, if there truly is no difference in that pair among the respondents, the result will average out to 50:50. The evaluation task of the subject is, thus, easier as compared to rating on a scale. However, if n products have to be evaluated, then the subject is required to go through each of the n(n-1)/2 possible number of pairs and identify the better product in each pair. Thus, if 5 products need to be evaluated, then the number of possible pairs would be 5(5-1)/2 = 10.

The major advantage of the paired comparison approach is that it makes the subject's tasks simple and more accurate as the subject has to only compare the two products in each trial and only identify the better product in the pair. The disadvan-tage of the paired comparison approach is that as the number of products (n) to be evaluated increases, the number of possible paired comparison judgments that each subject needs to make increases rapidly (proportional to the square of n) and the entire evaluation process becomes very time-consuming.

We will review application procedures for two commonly used methods based on the paired comparison approach, a) Thurstone's Method of Paired Comparisons, and b) Analytical Hierarchical Method. Thurstone's method allows us to develop scale values for each of the n products on a z-scale (z is a normally distributed vari-able with mean equal to zero and standard deviation equal to one) of desirability (Thurstone, 1927); whereas the Analytical Hierarchical Method allows us to obtain relative importance weights of each of the *n* products (Satty, 1980). Both the methods are simple and quick to administer and have the potential of providing more reliable evaluation results as compared to other subjective methods where a subject is asked to evaluate one product at a time.

Thurstone's Method of Paired Comparisons

Let us assume that we have five products (or designs or issues) that need to be evaluated. The five products are named: S, W, N, P and K. The 10 possible pairs of the product are: 1) S and W; 2) S and N; 3) S and P; 4) S and K; 5) W and N; 6) W and P; 7) W and K; 8) N and P; 9) N and K; and 10) P and K. The steps to be used in the procedure are presented below.

Step 1: Select an Attribute for Evaluation of the Products
The purpose of the evaluation is to order five products along an interval scale based on a selected attribute. Let us assume that the five products are outside door handles used to open the driver's door. The five designs are assumed to differ due to the shape

of their grasp areas and operating movements. The attribute selected is "ease of operation of the door handle during door opening".

Step 2: Prepare the Products for Evaluations
It is further assumed that for the evaluations, five identical doors have been built and mounted in five identical vehicle bodies. Each door is fitted with one of the five door handles with their latches and latching mechanisms. The five vehicle bodies will be positioned at the same height and orientation in a test area with respect to the evaluator.

Step 3. Obtain Responses of Each Subject on All Pairs
It is also assumed that eighty subjects will be selected randomly from the population of the likely owners of the vehicle for the evaluation study. Each subject will be brought to the test area separately by an experimenter. The experimenter will provide instructions to the subject and ask the subject to open and close each door within each selected pair of doors and ask the subject to select the door handle that is easier to open in each pair. The pairs of doors will be presented in a random order to each subject and the random order will be different for each subject.

The responses of an individual subject are illustrated in Table 19.4. Each cell of the table presents "Yes" or "No" depending upon if the handle shown in the column was better (easier to open) than the handle shown in the row. It should be noted that only the 10 cells above the diagonal (marked by x) need to be evaluated.

Step 4. Summarize Responses of All Subjects in Terms of Proportion of Product in the Column Better than the Product in the Row
After all the subjects have provided responses, the responses are summarized as shown in Table 19.5 by assigning a "1" to a "Yes" response, and a "0" to a "No" response. Thus, the cell corresponding to "W" column and "S" row indicates that only 1 out the 80 subjects judged the handle "W" to be better (easier to operate) than handle "S".

The complements of the summarized ratings in Table 19.5 are entered in the cells below the diagonal as shown in Table 19.6. For example, the complement of "1/80 responses of product W better than product S" is "79/80 responses of product S better

---

TABLE 19.4
Responses of an Individual Subject for the Ten Possible Product Pairs

	S	W	N	P	K
S	x	No	No	No	No
W		x	No	No	Yes
N			x	No	Yes
P				x	Yes
K					x

Note: A "Yes" response indicates that the product shown in the column is better than the product in the row. A "No" response indicates that the product shown in the row was better than the product shown in the column.

TABLE 19.5
Number of Subjects Preferring Product in the Column over the Product in the Row Divided by Number of Subjects

	S	W	N	P	K
S	x	1/80	3/80	2/80	4/80
W		x	3/80	30/80	50/80
N			x	30/80	50/80
P				x	60/80
K					x

TABLE 19.6
Response Ratio Matrix with Lower Half of the Matrix Filled with Complementary Ratios

	S	W	N	P	K
S	x	1/80	3/80	2/80	4/80
W	79/80	x	3/80	30/80	50/80
N	77/80	77/80	x	30/80	50/80
P	78/80	50/80	50/80	x	60/80
K	76/80	30/80	30/80	20/80	x

TABLE 19.7
Proportion of Preferred Responses (p_{ij})

		i=1	i=2	i=3	i=4	i=5
		S	W	N	P	K
j=1	S	x	0.013	0.038	0.025	0.050
j=2	W	0.988	x	0.038	0.375	0.625
j=3	N	0.963	0.963	x	0.375	0.625
j=4	P	0.975	0.625	0.625	x	0.750
j=5	K	0.950	0.375	0.375	0.250	x

than product W". The proportions in Table 19.6 are expressed in decimals in Table 19.7. Each cell in the matrix presented in Table 19.7 thus represents proportion p_{ij} indicating the proportion of responses in which the product in the ith column was preferred over the product in the jth row.

Step 5: Adjusting p_{ij} Values
To avoid the problem of distorting the scale values (computed in next step) of the products when p_{ij} values are very small (close to 0.00, or to 1.00), the proportion

values in Table 19.7 above 0.977 are set to 0.977 and the proportion values below 0.023 are set to 0.023 as shown in Table 19.8. (These adjustments are done to avoid the z-values in the next step straying further into the tails of the normal distribution and distorting the scale values.)

Step 6: Computation of Z-values and Scale Values for the Products
In this step, the values of the proportions (p_{ij}) in each cell are converted into Z-values by using the table of standardized normal distribution found in any standard statistics textbook. For example, the value of $p_{21} = 0.023$ is obtained by integrating the area under the standardized normal distribution curve (with mean equal to 0 and standard deviation equal to 1.0) from minus infinity to -1.995. Thus, a Z-value of -1.995 provides a p-value of 0.023. The Z-values can also be obtained by using a function called NORMINV by setting its parameters as $(p_{ij}, 0, 1)$ in Microsoft Excel. The Z-values (Z_{ij}) are obtained by converting all the proportion (p_{ij}) values in Table 19.8 by using the above conversion procedure are shown in the matrix on the top part of Table 19.9.

TABLE 19.8
Adjusted Table of p_{ij} (If p_{ij} > 0.977, then set p_{ij} = 0.977; and if p_{ij} < 0.023, then set p_{ij} = 0.023)

		i=1	i=2	i=3	i=4	i=5
		S	W	N	P	K
j=1	S	x	0.023	0.038	0.025	0.050
j=2	W	0.977	x	0.038	0.375	0.625
j=3	N	0.963	0.963	x	0.375	0.625
j=4	P	0.975	0.625	0.625	x	0.750
j=5	K	0.950	0.375	0.375	0.250	x

TABLE 19.9
Values of Z_{ij} Corresponding to Each p_{ij} and Computation of Scale Values (S_j)

		i=1	i=2	i=3	i=4	i=5
		S	W	N	P	K
j=1	S	x	-1.995	-1.780	-1.960	-1.645
j=2	W	1.995	x	-1.780	-0.319	0.319
j=3	N	1.780	1.780	x	-0.319	0.319
j=4	P	1.960	0.319	0.319	x	0.674
j=5	K	1.645	-0.319	-0.319	-0.674	x
	ΣZ_{ij} =	7.381	-0.215	-3.561	-3.272	-0.333
	S_j =	2.088	-0.061	-1.007	-0.925	-0.094

Note: Z_{ij} = Value of NORMINV(p_{ij},0,1) function from the Microsoft Excel.

The Z-values obtained in each column are summed, and the scale values for each product (S_i) are obtained by using the following formula (see last two rows of Table 19.9):

$$S_i = (\sqrt{2/n}) \Sigma Z_{ij}$$

Where n = number of products used in paired comparisons

The bottom row of Table 19.9 presents the scale values (S_i) for each product. (Note: using n=5 in the above formula). It should be noted that the sum of the scale values computed from the above formula is equal to 0.0 (i.e., $\Sigma S_i = 0.0$).

Figure 19.2 presents a bar chart of the scale values (S_i) of the five products shown in Table 19.9. Thus, the above procedure shows that by using Thurstone's method of paired comparisons scale values of the products are obtained. The scale values indicate the strength of the relative preference of each of the products in the set of the n products. The unit of the scale values is in number of standard deviations and the "zero" value on the scale corresponds to the point of indifference (i.e., the product with the zero scale value is neither liked (preferred) nor disliked (not preferred). Thus, in this example, product S is the best (most preferred) among the five products and product N is least preferred.

Analytical Hierarchical Method

In the analytical hierarchical method (AHM), the products are also compared in pairs. However, the better product in each pair is also rated in terms of the strength of the attribute it possesses in relation to the strength of the same attribute in the other product in the pair. The strength of the attribute is expressed using a ratio scale. The scale (or the weight) value of 1 is used to denote equal strength of the attribute in both

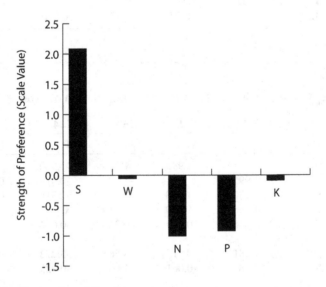

FIGURE 19.2   Scale values of the five products.

the products in the pair. And the scale value of 9 is used to indicate extreme or absolute strength of the attribute in the better product. And the product with the weaker strength is assigned the inverse of the scale value of the better product. The following example will illustrate this rating procedure.

Let us assume that there are two products U and R in a pair and the attribute to compare the products is "ease of use". The scale values assigned to the products using the ratio scale would be as follows:

1) If product U is "extremely or absolutely easy" to use as compared to product R, then, the weight of U preferred over R will be 9, and the weight of R preferred over U will be 1/9.

2) If product U is "very easy" to use as compared to product R, then, the weight of U preferred over R will be 7, and the weight of R preferred over U will be 1/7.

3) If product U is "easy" to use as compared product R, then, the weight of U preferred over R will be 5, and the weight of R preferred over U will be 1/5.

4) If product U is "moderately easy" to use as compared to product R, then, the weight of U preferred over R will be 3, and the weight of R preferred over U will be 1/3.

5) If product U is "equally easy" to use as compared to product R, then, the weight of U preferred over R will be 1, and the weight of R preferred over U will also be 1.

When a decision maker compares two items in a pair for a weight of importance (or preference), Satty (1980) described the 9-point scale by using the following adjectives.

1 = Equal importance
2 = Weak importance
3 = Moderate importance
4 = Moderate plus importance
5 = Strong importance
6 = Strong plus importance
7 = Very strong or demonstrated importance
8 = Very, very strong importance
9 = Extreme or absolute importance

From the viewpoint of making the scales more understandable, usually only the odd numbered scale values (shown in bold case above) are described and presented to the subjects. To allow the subjects to decide on the weight, the author found that the scale presented in Figure 19.3 works very well. Here the subject will be asked to put an "X" mark on the scale on the left side if product U is preferable over R. The higher numbers on the scale indicate higher preference. If both products are equally preferred, then the subject will be asked to place the "X" mark at the mid-point of scale value equal to 1. If product R is preferred over product U, then the subject will use the right side of the scale.

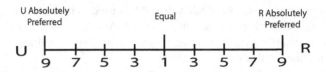

FIGURE 19.3 Scale used to indicate strength of the preference when comparing two products (U and R).

TABLE 19.10
Matrix of Paired Comparison Responses for One Evaluator

	U	R	T	M	L	P
U	1	5/1	1/1	7/1	1/1	1/1
R	1/5	1	1/2	5/1	1/1	3/1
T	1/1	2/1	1	3/1	5/1	3/1
M	1/7	1/5	1/3	1	1/1	1/3
L	1/1	1/1	1/5	1/1	1	1/3
P	1/1	1/3	1/3	3/1	3/1	1

Note: The value in a cell indicates the preference ratio for comparing the product in a row with the product in a column.

Let us assume that we have to compare six products, namely, U, R, T, M, L and P by using the analytical hierarchical technique. A subject will be asked to compare the products in pairs. The 15 possible pairs of the 6 products will be presented to the subject in a random order. The subject will be given a preselected attribute (e.g., ease of use) and asked to provide strength of preference ratings for each of the 15 pairs by using scales such the one presented in Figure 19.3. The data obtained from the 15 pairs will then be tabulated into a matrix of paired comparison responses as shown in Table 19.10. Each cell of the matrix indicates the ratio of preference weight of the product in the row over the product in the column. Thus, the ratio 5/1 in the first row and second column indicates that the product in row (U) was preferred (i.e., considered to be easy with rating weight of 5) over the product in column (R).

To compute the relative weights of importance of the products, the fractional values in Table 19.10 are first converted into decimal numbers as shown in the left side matrix in Table 19.11. All the six values in each row are then multiplied together and entered in the column labeled as "Row Product" in Table 19.11. The geometric mean of each row product is computed. It should be noted that the geometric mean of the product of n numbers is the $(1/n)^{th}$ root of the product (e.g., 1/6th root of 35.00 is $35^{(1/6)} = 1.8086$). All the six geometric means in the column labeled as "Geometric Mean" are then added. The sum, as shown in Table 19.11, is 7.0099. Each of the geometric means is then divided by their sum (7.0099) to obtain the normalized weight of the products. It should be noted that due to the normalization, the sum of the normalized weights over all the products is 1.0.

TABLE 19.11
Computation of Normalized Weights of the Product Attribute

	U	R	T	M	L	P	Row Product	Geometric Mean	Normalized Column
							35.0000	1.8086	0.2580
U	1.00	5.00	1.00	7.00	1.00	1.00	1.5000	1.0699	0.1526
R	0.20	1.00	0.50	5.00	1.00	3.00	90.0000	2.1169	0.3020
T	1.00	2.00	1.00	3.00	5.00	3.00	0.0031	0.3821	0.0545
M	0.14	0.20	0.33	1.00	1.00	0.33	0.0660	0.6357	0.0907
L	1.00	1.00	0.20	1.00	1.00	0.33	0.9801	0.9967	0.1422
P	1.00	0.33	0.33	3.00	3.00	1.00	Sum-->	7.0099	1.000

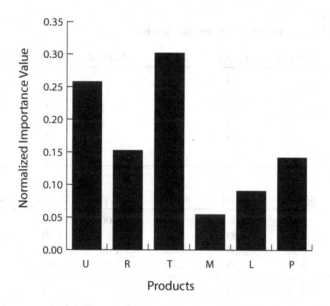

FIGURE 19.4 Normalized weights of the six products.

The normalized weights are plotted in Figure 19.4. The figure, thus, shows that the most preferred product (based on the ease of use) was T, and the least preferred product was M. The above example was based on data obtained from one subject. If more subjects are available, then normalized weights for each subject can be obtained by using the above procedure and then average weights of each product can obtained by averaging over the normalized weights of all the subjects for each product.

APPLICATION OF ANALYTICAL HIERARCHY METHOD TO A MULTI-ATTRIBUTES PROBLEM

The application of AHM for evaluating a set of items (or alternatives) using multiple attributes involves a three-step procedure. In step 1, the weights of attributes are determined by applying the AHM described in the preceding section. Then in step 2, weights the alternatives are determined for each attribute separately using the AHM. And in step 3, the weights obtained for the attributes in step 1 and weights of alternatives obtained for each of the attributes in step 2 are used to obtain the final weights of the alternatives.

The three steps are illustrated in Tables 19.12, 19.13 and 19.14 for steps 1, 2 and 3, respectively. The problem illustrated here involves determination of weights of five alternatives (i.e., five types of vehicles) based on six attributes (dimensions)

TABLE 19.12
Step 1: Obtaining Weights for the Six Attributes

	SgRPHt	EntHt	RocHt	RocWd	StoRoc	MaptoRoc
SgRPHt	1	5	3	3	5	7
EntHt	1/5	1	1/3	1/3	5	5
RocHt	1/3	3	1	1	5	5
RocWd	1/3	3	1	1	3	5
StoRoc	1/5	1/5	1/5	1/3	1	3
MaptoRoc	1/7	1/5	1/5	1/5	1/3	1

	SgRPHt	EntHt	RocHt	RocWd	StoRoc	MaptoRoc
SgRPHt	1.00	5.00	3.00	3.00	5.00	7.00
EntHt	0.20	1.00	0.33	0.33	5.00	5.00
RocHt	0.33	3.00	1.00	1.00	5.00	5.00
RocWd	0.33	3.00	1.00	1.00	3.00	5.00
StoRoc	0.20	0.20	0.20	0.33	1.00	3.00
MaptoRoc	0.14	0.20	0.20	0.20	0.33	1.00

Row Product	Geometric Mean	Normalized Weight
1575.0000	3.4110	0.410292482
0.5544	0.9064	0.109024276
24.9750	1.7097	0.2056511058
14.9850	1.5702	0.1888670763
0.0080	0.4471	0.053784365
0.0004	0.2692	0.0323806948
Sum-->	8.3136	1.000

TABLE 19.13
Step 2: Obtaining Weights for Alternatives for Each Attribute

A. SgRP Height from the ground (H5) | **Preferred: SgRP height near standing buttock height**

Importance of criterion in row over criterion in column

	Fulpickup	MidSUV	MPV	LgSedan	SpoCar	Row Product	Geometric Mean	Normalized Weight
Fulpickup	1.00	0.20	0.14	0.33	1.00	0.0095	0.5589	0.0985
MidSUV	5.00	1.00	1.00	3.00	5.00	75.0000	1.7155	0.3022
MPV	7.00	1.00	1.00	3.00	5.00	105.0000	1.7892	0.3152
LgSedan	3.00	0.33	0.33	1.00	5.00	1.6683	1.0661	0.1878
SpoCar	1.00	0.20	0.20	0.20	1.00	0.0080	0.5469	0.0963
						Sum-->	5.6764	1.000

B. Entrance height (H11) | **Preferred: Taller entrance height**

	Fulpickup	MidSUV	MPV	LgSedan	SpoCar	Row Product	Geometric Mean	Normalized Weight
Fulpickup	1.00	7.00	7.00	9.00	9.00	3969.0000	2.8173	0.4536
MidSUV	0.14	1.00	3.00	3.00	7.00	9.0000	1.3161	0.2119
MPV	0.14	0.33	1.00	1.00	5.00	0.2381	0.8358	0.1346
LgSedan	0.11	0.33	1.00	1.00	9.00	0.3333	0.8717	0.1403
SpoCar	0.11	0.14	0.20	0.11	1.00	0.0004	0.3702	0.0596
						Sum-->	6.2111	1.000

C. Rocker Height (G) | **Preferred: Lower rocker height**

	Fulpickup	MidSUV	MPV	LgSedan	SpoCar	Row Product	Geometric Mean	Normalized Weight
Fulpickup	1.00	0.20	0.14	0.14	3.00	0.0122	0.5768	0.0978
MidSUV	5.00	1.00	0.33	3.00	7.00	35.0000	1.5596	0.2643
MPV	7.00	3.00	1.00	3.00	7.00	441.0000	2.1407	0.3628
LgSedan	7.00	0.33	0.33	1.00	5.00	3.8889	1.1850	0.2008
SpoCar	0.33	0.14	0.14	0.20	1.00	0.0014	0.4382	0.0743
						Sum-->	5.9003	1.000

D. Rocker width (W) | **Preferred: Shorter rocker width**

	Fulpickup	MidSUV	MPV	LgSedan	SpoCar	Row Product	Geometric Mean	Normalized Weight
Fulpickup	1.00	0.40	0.14	0.20	0.11	0.0013	0.4345	0.0792
MidSUV	2.50	1.00	0.33	0.33	3.00	0.8333	0.9775	0.1782
MPV	7.00	3.00	1.00	1.00	1.00	21.0000	1.4631	0.2668
LgSedan	5.00	3.00	1.00	1.00	3.00	45.0000	1.6094	0.2934
SpoCar	9.00	0.33	1.00	0.33	1.00	1.0000	1.0000	0.1823
						Sum-->	5.4844	1.000

(continued)

TABLE 19.13 (Continued)
Step 2: Obtaining Weights for Alternatives for Each Attribute

E. Seat edge to rocker (S)   **Preferred: Shorter rocker width**

	Fulpickup	MidSUV	MPV	LgSedan	SpoCar	Row Product	Geometric Mean	Normalized Weight
Fulpickup	1.00	0.33	0.20	0.33	0.20	0.0044	0.5081	0.0952
MidSUV	3.00	1.00	0.33	1.00	1.00	1.0000	1.0000	0.1873
MPV	5.00	3.00	1.00	3.00	1.00	45.0000	1.6094	0.3014
LgSedan	3.00	1.00	0.33	1.00	1.00	1.0000	1.0000	0.1873
SpoCar	5.00	1.00	1.00	1.00	1.00	5.0000	1.2228	0.2290
						Sum-->	5.3403	1.000

F. Map pocket to rocker (T)   **Preferred: Short pappocker to rocker distance**

	Fulpickup	MidSUV	MPV	LgSedan	SpoCar	Row Product	Geometric Mean	Normalized Weight
Fulpickup	1.00	0.20	0.33	0.33	0.20	0.0044	0.5081	0.0920
MidSUV	5.00	1.00	0.33	0.33	0.33	0.1852	0.8099	0.1467
MPV	3.00	3.00	1.00	3.00	0.20	5.4000	1.2347	0.2236
LgSedan	3.00	3.00	0.33	1.00	0.33	1.0000	1.0000	0.1811
SpoCar	5.00	3.00	5.00	3.00	1.00	225.0000	1.9680	0.3565
						Sum-->	5.5207	1.000

of the vehicles related to the driver's entry/exit performance (see Volume 1, Chapter 10).

The five alternatives (vehicle types) are as follows:

1) Full-size pickup truck [Fulpickup](e.g., Ford F-150, Chevy Silverado)
2) Mid-size SUV [MidSUV](e.g., Ford Explorer, Chevy Traverse)
3) Mid-size passenger van [MPV](e.g., Toyota Sienna, Honda Odyssey)
4) Large Sedan [LgSedan](e.g., Ford Taurus, Toyota Camry)
5) Sports car [SpoCar](e.g., Mazda Miata, Chevy Corvette)

The six attributes (dimensions) of the vehicles considered for evaluating the above eight alternatives are:

1) SgRPHt = Height of SgRP from the ground (H5)
2) EntHt = Entrance height (H11)
3) RocHt = Rocker height (G)
4) RocWd = Rocker width (W)
5) StoRoc = Seat edge to rocker (S)
6) MaptoRoc = Map pocket to rocker (T)

The vehicle dimensions shown above are illustrated in Volume 1, Figure 10.2.

TABLE 19.14
Step 3: Obtaining Final Weights for Alternatives

Attributes -->		SgRPHt	EntHt	RocHt	RocWd	StoRoc	MaptoRoc	Final Weights
		Final weighting of scores:						
Normalized Weights-->		0.4103	0.1090	0.2057	0.1889	0.0538	0.0324	
Normalized Weights for Vehicle Types	Fulpickup	0.0985	0.4536	0.0978	0.0792	0.0952	0.0920	0.1330
	MidSUV	0.3022	0.2119	0.2643	0.1782	0.1873	0.1467	0.2499
	MPV	0.3152	0.1346	0.3628	0.2668	0.3014	0.2236	0.2924
	LgSedan	0.1878	0.1403	0.2008	0.2934	0.1873	0.1811	0.2050
	SpoCar	0.0963	0.0596	0.0743	0.1823	0.2290	0.3565	0.1196
SUM-->		1.0000	1.0000	1.0000	1.0000	1.0000	1.0000	1.0000

Importance of criterion in row over criterion in column Preferred: Safe during accidents

The final weights of the vehicles are shown in Table 19.14.

The weight for full-size pickup (Fulpickup) is computed as follows:

(0.4103x0.0985) + (0.1090x0.4536) + (0.2057x0.0978) + (0.1889x0.0792) + (0.0538x0.0952) + (0.0324x0.0920) = 0.1330

Similarly, the final weights of other vehicle types are computed by obtaining a sum of multiplications over all attributes. Each multiplication (for each attribute) is computed by multiplying the normalized weight of each attribute by normalized weight of each vehicle type for the same attribute.

The last column in Table 19.14 shows that MPV obtained the highest weight of 0.2924 and the Spocar obtained the lowest weight of 0.1196.

When more than one subject (or expert) is available, the above described three-step process should be applied by using each subject (or expert) separately. The final weights of alternatives obtained for each alternative in step 3 for each subject can be aggregated by computing arithmetic (or geometric) means of the weights of the alternatives obtained for each subject.

CONCLUDING REMARKS

Evaluations of the product is a continuous process from development of early concepts to the final product after it is purchased and used by the customer. Thus, every entity within the product from component to the assembled whole product undergoes evaluations to ensure that every attribute requirement is met. The evaluation methods for each attribute and its lower level attributes are unique depending upon the level of entities within the product available for testing. Each chapter of this book deals with ergonomic considerations in specialized areas such as occupant packaging, field of view, entry/exit, and lighting systems and each area requires many evaluations that are described in corresponding chapters. The evaluation needs and methods were summarized in Table 19.1. Ergonomics engineers should be knowledgeable about the evaluation tests being conducted by different departments and provide additional inputs to ensure that ergonomic considerations are included during these evaluations.

REFERENCES

Bertollini, G., L. Brainer, J. Chestnut, S. Oja and J. Szczerba. 2010. *General Motors Driving Simulator and Applications to Human Machine Interface (HMI) Development.* SAE paper no. 2010-01-1037. Presented at the 2009 SAE World Congress, Detroit, MI.

Besterfield, D. H., C. Besterfield-Michna, G. H. Besterfield and M. Besterfield-Scare. 2003. *Total Quality Management.* ISBN: 0-13-099306-9, Third Edition. Upper Saddle River, NJ: Prentice Hall.

Bhise, V., V. Sarma and P. Mallick. 2009. *Determining Perceptual Characteristics of Automotive Interior Materials.* SAE Paper no. 2009-01-0017. Presented at the 2009 SAE World Congress, Detroit, MI.

Bhise, V., R. Hammoudeh, J. Dowd and M. Hayes. 2005. Understanding Customer Needs in Designing Automotive Center Consoles. *Proceedings of the Annual Meeting of the Human Factors and Ergonomics Society*, Orlando, FL.

Bhise, V., R. Boufelliga, T. Roney, J. Dowd and M. Hayes. 2006. *Development of Innovative Design Concepts for Automotive Center Consoles*. SAE Paper no. 2006-01-1474. Presented at the SAE 2006 World Congress, Detroit, MI.

Bhise, V., S. Onkar, M. Hayes, J. Dalpizzol, and J. Dowd. 2008. Touch Feel and Appearance Characteristics of Automotive Door Armrest Material. *Journal of Passenger Cars – Mechanical Systems*, SAE 2007 Transactions.

Bhise, V. D. 2023. *Designing Complex Products with Systems Engineering Processes and Techniques*. Second Edition. ISBN: 978-1-032-20369-0. Boca Raton, FL: CRC Press.

Bodenmiller, F., J. Hart and V. Bhise. 2002. *Effect of Vehicle Body Style on Vehicle Entry/Exit Performance and Preferences of Older and Younger Drivers*. SAE Paper no. 2002-01-00911. Paper presented at the SAE International Congress in Detroit, MI.

Chapanis, A. 1959. *Research Techniques in Human Engineering*. Baltimore, MD: The Johns Hopkins Press.

Creveling, C. M., J. L. Slutsky and D. Antis, Jr. 2003. *Design for Six Sigma – In Technology and Product Development*. Upper Saddle River, NJ: Prentice Hall PTR.

Jack, D. D., S. M. O'Day and V. D. Bhise. 1995. *Headlight Beam Pattern Evaluation – Customer to Engineer to Customer – A Continuation*. SAE paper no. 950592. Presented at the 1995 SAE International Congress, Detroit, MI., MA

Kolarik, W. J. 1995. *Creating Quality – Concepts, Systems, Strategies, and Tools*. New York, NY: McGraw-Hill.

Owens, J. M., S. B. McLaughlin and J. Sudweeks. 2010. *On-Road Comparison of Driving Performance Measures When Using Handheld and Voice-Control Interfaces for Cell Phones and MP3 Players*. SAE Paper no. 2010-01-1036. Presented at the 2010 SAE World Congress held in Detroit, MI.

Pew, R. W. 1993. *Experimental Design Methodology Assessment*. BBN Report No. 7917, Cambridge, MA: Bolt Beranek & Newman, Inc.

Richards, A. and V. Bhise. 2004. Evaluation of the PVM Methodology to Evaluate Vehicle Interior Packages. SAE Paper no. 2004-01-0370. Also published in SAE Report SP-1877, SAE International, Inc., Warrendale, PA.

Satty, T. L.1980. *The Analytic Hierarchy Process*. New York, NY: McGraw Hill.

Thurstone, L. L. 1927. The Method of Paired Comparisons for Social Values. *Journal of Abnormal and Social Psychology*, 21: 384–400.

Tijerina, L, E. Parmer and M. J. Goodman.1999. Driver Workload Assessment of Route Guidance System Destination Entry While Driving: A Test Track Study. Transportation Research Center, East Liberty, Ohio.

Yang, K. and B. El-Haik. 2003. *Design for Six Sigma – A Roadmap for Product Development*. New York, NY: McGraw-Hill.

Zikmund, W. G. and B. J. Babin. 2009. *Exploring Market Research*. Ninth Edition. Boston, MA: Cengage Learning. ISBN-10: 0-324-78844-4

20 Understanding Interfaces Between Vehicle Systems

INTRODUCTION

An automotive product contains many vehicle systems. The vehicle systems must be interfaced with other vehicle systems such that the systems work together to perform all functions of the vehicle. The vehicle systems must also interface with the human operators (i.e., driver, passenger and assembly, repair, maintenance personnel) to ensure that they have adequate spaces (with clearances) to perform their tasks. The automotive designers work with studio engineers and package engineers to create exterior and interior surfaces to form envelopes. All vehicle systems are packaged within their respective envelopes. In order for the systems to work with other systems, interfaces (i.e., connections) between the systems must be designed to ensure that all systems fit within the vehicle space and perform their allocated functions. In this chapter, we will review types of interfaces, interface diagrams and interface matrices used to understand interface design tasks and requirements on the interfaces.

INTERFACES

WHAT IS AN INTERFACE?

An interface can be defined as a "joint" (or connection) where two (or more) entities (e.g., systems, subsystems or components) are linked together to serve their allocated functions. Thus, the interface affects the design of both the entities and the parameters defining the joint (i.e., configuration of connecting elements at the interface). The joint or the interface between the two entities must be compatible, that is, the values of the parameters (e.g., dimensions of the interfacing portions and their capacities, electrical and signal communication characteristics) of the two interfacing entities defining their capabilities must match. An interface can involve a) physical connection (or attachment), b) sharing of space (i.e., packaged close to each other), c) exchange of energy (e.g., transfer of mechanical, hydraulic, electric, thermal or luminous energy), d) exchange of material (e.g., oil, coolant, gases), and/or, e) exchange of data (e.g., digital and/or analog signals).

Knowing the type of interface and its characteristics is important to ensure that the two interfacing entities work with each other to perform their allocated functions.

DOI: 10.1201/9781003485605-7

During the early design phases of the product, as the functions and their requirements are allocated and the systems are identified, the interfaces between different entities and their parameters must be also identified. As the design progresses further, the parameters that define each interface in terms of its characteristics (e.g., their dimensions, strength of physical attachment forces, amount of current or data flow passing through the interface), and their level of strength or capacity must be determined, configured and controlled during subsequent detailed design activities. The engineers involved in designing of both the interfacing entities must know how the two entities work with each other, and how and what the interface must exchange, communicate or share to get the two entities to work together and perform their intended functions.

It should be realized that since each system in a product performs one or more functions, all systems in a product, including the human operators – who can be considered as system components – must work together for the product to function. Thus, each interface must be carefully designed to ensure that both interfacing systems are compatible.

TYPES OF INTERFACES

Interfaces between systems, subsystems or components of a product and other external systems that affect the operation of the product and their components (e.g., parts, subassemblies, human operators, software) need to be studied and designed to ensure that the product can be used by its customers. Interfaces can be categorized by considering many engineering characteristics and user needs of the product (Lalli et al., 1997). Some commonly considered types of interfaces are described below.

1) *Mechanical or Physical Interface*: This type of interface ensures that any two interfacing components perform the following: a) they can be physically joined together (e.g., by use of bolts, rivets, threads, couplings, welds, or adhesives), b) their linkages (or joints) that are fixed (not movable) or allow range of movements (e.g., through pins or hinges), c) they can transmit forces between entities such as a link, spring, damper, or frictional element (e.g., the interface between brake drum and brake shoe pad), and d) they have the required strength or transfer capabilities (e.g., for transfer of materials, heat, or forces) and durability – that is, the ability to function under many work cycles involving loads, vibrations, temperatures, and so forth.

2) *Fluidic or Material Transfer Interface*: A fluidic or material transfer interface (for transfer of fluids, gases or powdered /granular materials) can be considered as a different type of interface, or it can be considered as a mechanical interface involving pipes, tubes, hoses, ducts, seals, and so forth. The fluidic interface enables the flow of fluids, gases or powdered/granular materials with their characteristics such as flow rates, purity, pressures, temperatures, insulation, sealing, corrosion resistance, and so forth.

3) *Packaging Space for Interfacing Entities*: Physical space is required to package or to accommodate the two interfacing entities. The required space

can be determined from: a) the sizes/volumes and shape of spaces (i.e., three-dimensional envelopes) occupied by the two interfacing entities and their interfaces, b) clearance spaces required around the entities to account for vibrations, movements of parts/linkages, air passages for cooling, hand/finger or tool access space for setting controls, assembly, service and repair, and c) consideration of minimum and maximum separation distances required for their operations (e.g., space between a fan and a heat exchanger). The packaging interface is also called the spatial interface. Some examples of packaging interfaces are a) engine is packaged within the engine compartment, b) the engine and the radiator are packaged together, and c) occupants are packaged within the passenger compartment.

4) *Functional Interface*: In some cases, depending upon certain needs to provide one or more functions, one or more of the above types of interfaces may be combined and defined as a functional interface. For example, an automotive suspension system forms a unique functional interface (involving physical links and their relationships with relative movements) between sprung and un-sprung masses of the vehicle.

5) *Electrical Interface*: An electric interface ensures that two interfacing entities can form an electrical connection/coupling (e.g., with connectors, pins, screws, soldering, and spring-loaded contacts/brushes) that can carry required electrical current or signals, provide necessary insulation protection, data transfer, and may have other characteristics such as resistance, capacitance, electro-magnetic fields, and interferences.

6) *Software Interface*: A software interface ensures that data are transferred from one entity (with a software system) to another. The format and transmission characteristics of the coded data through the two interfacing entities must be compatible to facilitate the required type, amount and rate of the data transfer.

7) *Magnetic Interface*: A magnetic interface generates the required magnetic fields for operation of devices such as solenoids/relays, electric machines (motors, generators) and levitation devices.

8) *Optical Interface*: An optical interface (e.g., fiber optics, light paths, light guides, light piping, mirrors or reflecting surfaces, lenses, prisms, and filters) allows transfer of light energy between adjoining entities through luminous or non-luminous (e.g., infrared) energy transmission and reflection. The interface may also function to prevent the radiant energy transfer by shielding, baffling/blocking, or filtering of unwanted radiated energy.

9) *Wireless Interface*: This type of interface can communicate signals or data without wires via radio frequency communication, microwave communication (e.g., long-range line-of-sight via highly directional antennas, or short-range communication), infrared (IR) short-range communication, Bluetooth, and so forth. The interface applications may involve point-to-point communication, point-to-multipoint communication, broadcasting, cellular networks and other wireless networks.

10) *Sensor or Actuator Interface*: A sensor has a unique interface that converts certain sensed energy (e.g., light, motion, touch, distance or proximity

to certain objects, pressure, and temperature) into an electrical output or signal. For example, a float or floating sensor device can sense fluid levels and convert into electrical signals. Whereas an actuator produces an output (e.g., movement of a control or mechanical links) by converting an input from one type of modality to a different output modality. For example, a stepper motor produces a precise angular movement for each electrical pulse input.

11) *Human Interface*: When a human operator is involved in operating, monitoring, controlling or maintaining a product, the human-machine or human-computer interface, commonly referred to as the HMI or HCI, respectively, will include devices such as human accommodating or positioning devices (e.g., chairs, seats, armrests, cockpits, standing platforms, steps, foot rests, handles, access doors), controls (e.g., steering wheel, gear shifter, switches, buttons, touch controls, stalks, levers, joy sticks, pedals, and voice controls), tools (e.g., hand tools, powered tools) and displays (e.g., visual displays, auditory displays, tactile displays and olfactory displays). During driving, the driver also gets feedback from various entities within the vehicles, such as vibrations from seats, steering wheel, pedals, tactile feedback from feel during control movements (e.g., feel during gear shifting), sensation of thermal comfort from operation of climate control system. The type of perception that the driver gets, for example, pleasant versus non-pleasant, quality versus shoddy, dependent on the characteristics of the interfaces and the entities within the vehicle.

INTERFACE REQUIREMENTS

In order to design an interface, an engineer must first understand the product's overall requirements, the allocated functions and characteristics of both the entities attached or linked at the interface. The requirements on the interface should specify the following: a) the functional performance of both the entities, b) configuration of the entities, c) available space to create the interface, d) environmental conditions for the operation of the product and comfort of the human operators, e) durability (minimum number of operational cycles the product must function), f) reliability and safety considerations in performing the required functions, g) human needs (e.g., viewing and reading needs, hearing needs (sound frequencies and levels), lighting and climate control needs and product operating needs), h) electro-magnetic interference, and so forth. In addition, the requirements should include any other special constraints, for example, weight requirements, aerodynamic considerations, and operating temperature ranges, that must be met.

Steps involved in the interface requirements development process generally use an iterative approach (with a series of steps and loops as shown in Volume 1, Figures 2.2 and 2.3) unless a previously developed requirements document (a "standard") is available. The series of steps typically involve the following:

1) Gather information to understand how the interfacing entities work, fit into the product and support the overall functionality, performance and requirements of the product (e.g., review existing system design documents and standards). Draw an interface diagram (described later in this chapter). Meet with the design team members of the interfacing entities for an automotive product – e.g., core engineering functions such as body engineering, powertrain engineering, electrical engineering, climate control engineering, and so on – and for product design teams to understand issues and trade-off considerations with the product attributes (e.g., packaging space, safety, maintenance, costs, and performance).

2) Document all design considerations such as inputs, outputs, constraints and trade-offs associated with the interface and its effects on other entities for example, develop a cause-and-effect diagram and conduct a FEMA (see Volume 1,Chapter 3 and Bhise, 2023).

3) Study existing designs of similar interfaces and compare them by benchmarking the competitors' products (see Volume 1, Chapter 3 for Benchmarking technique).

4) Study existing and new technologies that could be implemented to improve the interfaces.

5) Create an interface matrix (described in later in this chapter) of selected systems to understand all interfaces (between the selected systems and other vehicle systems), their types and their characteristics.

6) Create a preliminary set of requirements on how each interface should function.

7) Translate requirements into design specifications (use of the QFD technique can help in this step; see Volume 1, Chapter 3 and Bhise, 2023).

8) Brainstorm possible verification tests (or obtain available test methods from existing standards) that need to be performed to demonstrate compliance to the requirements.

9) Develop alternate interface concepts/ideas.

10) Review alternate concepts and ideas with subject matter technical experts (i.e., conduct design iterations; see Volume 1, Chapter 2).

11) Select a leading design by analyzing all other entities that are functionally linked to the entities associated with the interface (develop a Pugh diagram to aid in decision making; see Volume 1, Chapter 3).

12) Modify and refine interface diagram and interface matrix.

13) Iterate above steps until an acceptable interface design is found (see Volume 1, Chapter 2).

The iterative workload described in the above process can be reduced if an internal (company) design guide or standard for designing the entities being interfaced can be used as a starting document along with the product level requirements. Experts and other knowledgeable people in the organization can provide information on valuable lessons learned during the development of similar interfaces from past product programs.

VISUALIZING INTERFACES

REPRESENTING AN INTERFACE

An interface between any two entities (which could be systems, subsystems or components,) can be represented by use of a simple arrow diagram as shown in Figure 20.1.

The arrow (between the two entities) indicates a link (or relationship) between the two entities, namely entity A and entity B. The arrow representing the link can denote any of the following (see Figure 20.1):

a) Output of entity A is an input to entity B
b) Entity A is mechanically attached to entity B
c) Entity A is functionally attached to entity B (i.e., function of A is required by B to perform its function)
d) Entity A provides information to entity B
e) Entity A provides energy to entity B
f) Entity A transmits or sends signals, data or material (e.g., fluids, gases) to entity B

For example, in an automobile, the interior door trim panel (on which the door armrest and door switches are mounted) is physically attached (using plastic press-fit studs) to the sheet metal door frame, and the door frame in turn is physically mounted (bolted using screws) to the vehicle body via door hinges (see Figure 20.2). (Note: Letter "P" placed above the arrows in Figure 20.2 indicates "physical" connection).

INTERFACE DIAGRAM

An interface diagram is a flow diagram (or an arrow diagram) showing how different systems, subsystems and components of a product shown in blocks (or rectangles in

FIGURE 20.1 Interface between two entities.

FIGURE 20.2 Interfaces between the interior door trim panel and the vehicle body.

the flow diagram) are interfaced (i.e., joined or linked) by arrows. It provides a visual representation of the product or a portion of the product showing where the interfaces occur. It also should show the type of interface by use of letter codes such as, "P" for a physical connection, "E" for an energy transfer, "M" for material/fluid transfer, "D" for data transfer, and so forth, placed next to the arrow.

An interface diagram is a useful tool in understanding how various systems, subsystems and components are interfaced with each other. The diagram can be created at any level, that is, at the product (i.e., vehicle) level showing all the systems of the product, at a system level showing all the subsystems of the system, at a sub-system level showing all components of the subsystem, or at a mixed level showing a system, its subsystems and also showing other major systems of the product. Two examples of the interface diagram are shown in later sections of this chapter. Figure 20.4 presents an interface diagram of vehicle systems in a car and Figure 20.5 presents an interface diagram for an automotive braking system.

INTERFACE MATRIX

An interface matrix is a commonly used method to illustrate the existence and types of interfaces between different entities (i.e., systems, subsystems or components). All the entities involved in the analysis are represented in the matrix. The entities are shown as headings for both the rows and the columns of the matrix. The headings for the rows are placed on the left side of the matrix. And the headings for the columns are placed above the matrix. Each cell of matrix is defined by the intersection of its row and column represented by the two interfacing entities. The description of the interface is shown in the cell by one or more applicable letter codes to indicate the type(s) of the interface.

Figure 20.3 presents the output-to-input relationships between six entities in a 6x6 interface matrix (except for the cells in the diagonal). The entities are labeled as *E1* to *E6*. The outputs of the six entities are labeled as *O1* to *O6,* and the inputs are shown as *I1* to *I6.*The arrow shown in each cell indicates that the output of an entity defined by its row is used as an input to an entity defined by the column of the matrix. For example, the cell in the first row and second column shows that *O1* is the output of entity *E1;* and it is interfaced with entity *E2* shown as input *I2* received by entity *E2*.

The contents of the cells of the interface matrix typically present coded descriptors of the types of interfaces between the outputting entities and the entities receiving the inputs. The codes typically include P = physical interface, S = spatial-packaging interface, E= energy transfer, M= material flow, I = information or data flow, and 0 (or a blank cell) = no relationship.

An interface matrix, thus, shows the following a) it captures the existence of all interfaces, b) it shows output-to-input relationships between any two entities (see Figure 20.3), and c) it presents type(s) of interface between any two entities. Examples of interface matrices are provided in the next section (see Tables 20.1 and 20.2). The interface matrix is also called an interaction matrix in some organizations.

The interface diagram and interface matrix are both very useful tools in visual-izing relationships and documenting the presence of the interfaces (Scaka, 2008).

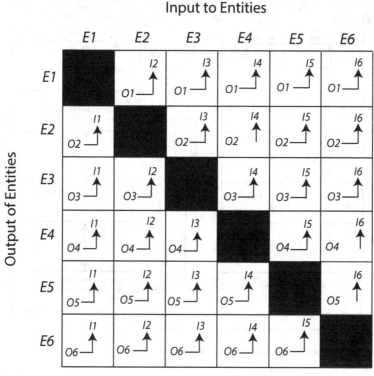

Input to Entities

FIGURE 20.3 Output-to-input relationships of entities indicated by the cells of the interface matrix.

These tools make the design team realize the presence of many interfaces, and the types of these interfaces in the product. The next step is to understand the connection configuration details, functional requirements of the interfacing entities and develop requirements of these interfaces to ensure that the interfacing entities can be designed to work together to perform their allocated functions.

EXAMPLES OF INTERFACE DIAGRAM AND INTERFACE MATRIX

Vehicle Systems Interface Diagram and Interface Matrix

Figure 20.4 illustrates an interface diagram for all major systems in a vehicle. All the eight major vehicle systems presented in Table 1.1 (Volume 1) are shown in the blocks in the interface diagram. The arrows between the blocks show interfaces between the

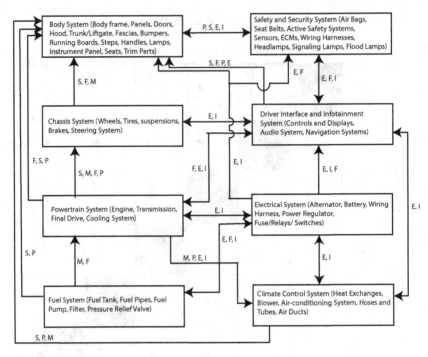

FIGURE 20.4   Interface diagram of vehicle systems.

systems and the letter codes above, or the right side of each arrow indicates the type of interface. The interface diagram thus shows that every system in the vehicle is attached to several other systems of the vehicle. For example, all vehicle systems are attached to the body system, which positions and holds all the systems to create the vehicle. The electrical system that provides the electrical power is also interfaced to all other vehicle systems.

Table 20.1 presents an interface matrix illustrating interfaces between all the major vehicle systems shown in Figure 20.4. The advantage of the interface matrix over the interface diagram is that it presents the interface information in an easy-to-follow format. One can look across each row to determine how the outputs of the system represented by the row are linked to other vehicle systems. For example, scanning across all columns and down all the rows, the matrix shows that body system, powertrain system, electrical system and driver interface system have the greatest number of interfaces to the other vehicle systems Thus, the engineers working on these systems must be in constant communication with other vehicle systems engineers to ensure that all the identified interfaces are designed to meet their respective requirements. Similarly, scanning horizontally across all columns indicates the interfaces that receive the inputs from the systems in the respective rows.

TABLE 20.1
Interface Matrix Illustrating Interfaces between All the Major Vehicle Systems

	Body System	Chassis System	Powertrain System	Fuel System	Electrical System	Climate Control System	Safety and Security System	Driver Interface and Infotainment System
Body System (Body frame, Panels, Doors, Hood, Trunk/Liftgate, Fascias, Bumpers, Running Boards, Steps, Handles, Lamps, Instrument Panel, Seats, Trim Parts)		S, F, M	P, S	P,S	P, S, E	P, S, E	P, S, E, I	P, S, E, I
Chassis System (Wheels, Tires, suspensions, Brakes, Steering System)	M, F, S		F, P, S	P, S,			E, I	E, I
Powertrain System (Engine, Transmission, Final Drive, Cooling System)	S, F	S, F, M		F, M	E, I	M, P, E, I	E, I	E, I
Fuel System (Fuel Tank, Fuel Pipes, Fuel Pump, Filter, Pressure Relief Valve)	S	P	M, F		F, E			I
Electrical System (Alternator, Battery, Wiring Harness, Power Regulator, Fuse/Relays/ Switches)	F, E, M	F	F, E	E		E, I	E, F	F, E, I
Climate Control System (Heat Exchanges, Blower, Air-conditioning System, Hoses and Tubes, Air Ducts)	M, F, S		F, P, E		E, I			E, I
Safety and Security System (Air Bags, Seat Belts, Active Safety Systems, Sensors, ECMs, Wiring Harnesses, Headlamps, Signaling Lamps, Flood Lamps)	M, S, I		F, E	I	E, I			E, F, I
Driver Interface and Infotainment System (Controls and Displays, Audio System, Navigation Systems)	M, S, I		F, E,I	I	E, I	E, I		

Notaion:

P = Physical interface
S = Spatial-packaging interface
E = Electrical interface
M = Material flow
I = Information or data flow
F = Functional interface

Driver Interfaces

In the interface diagram and interface matrix presented above, the driver can be shown as an additional entity. These interfaces can be designated as "DO" (driver output received as inputs to vehicle systems) and "DI" (input received by the driver from outputs of vehicle systems) to vehicle systems in an interface diagram and an interface matrix.

Some examples of driver interfaces to other vehicle systems are described below. Driver outputs providing inputs to the vehicle systems (DO):

1. Body system:
   a) The driver adjusts the seat controls to find preferred driving posture and positions in the seat.
   b) The pedals receive driver inputs for longitudinal control of the vehicle.
2. Chassis System
   a) The driver provides steering wheel movements for lateral control of the vehicle.
   b) The driver provides brake pedal force to decelerate the vehicle.
3. Powertrain System:
   a) The driver selects gear shift position.
   b) The driver provides accelerator and clutch pedal movements.
4. Fuel System:
   a) The driver fills gas by opening the fuel filler door, the fuel cap and inserting the fuel pump nozzle in the fuel filler tube.
   b) The driver reads the fuel gas to decide if refueling is needed.
5. Electrical System:
   a) Driver control movements affect operation of the electrical features.
   b) The driver pushes the ignition (starter) switch.
6. Climate Control Systems:
   a) The driver moves controls within the climate control module.
   b) The driver turns on the windshield defroster as the windshield begins to freeze.
7. Safety and Security System:
   a) The driver buckles the seat belt.
   b) The driver's vehicle locking action engages theft protection system.
   c) The driver operates headlamps and turn signal lamps.
8. Driver Interface and Entertainment System:
   a) The driver operates the audio controls to find a radio station.
   b) The driver reduces sound volume to hear the spoken words of a passenger.

Vehicle system outputs providing inputs to the driver (DI):

1. Body system:
   a) The seat supports the driver's body and, hence, holds the driver in position.

    b) The brake and accelerator pedals provide feedback in terms of pedal feel.

    c) The driver obtains feedback during control movements.

    d) Driver obtains visual information from the daylight (window) openings and mirror systems.

2. Chassis System

    a) The driver obtains tactile and vibratory feedback through a hand on the steering wheel and a foot on a pedal.

    b) The driver visually perceives change in vehicle heading (yawing), longitudinal and lateral motion, pitching and rolling of the of the vehicle.

3. Powertrain System:

    a) The driver obtains feedback from the engine and transmission (engine sound, shift feel).

    b) The driver feels maximum vehicle acceleration during 0 to 60 mph drive test.

4. Fuel System:

    a) The driver obtains fuel level information from the fuel gauge.

    b) The driver stops refueling after hearing the sound of fuel filling stoppage.

5. Electrical System:

    a) The driver obtains status of alternator/battery functions (charging/discharging) through warning lights and gauges.

    b) The driver hears the sound of the starter motor during engine start.

6. Climate Control Systems:

    a) Driver senses thermal comfort through air flow and air temperature.

    b) The driver hears the sound of rotating fan (blower) motor.

7. Safety and Security System:

    a) Vehicle displays provide status of safety and security systems.

    b) The driver hears the seatbelt reminder chime if the seatbelt is not fastened.

8. Driver Interface and Entertainment System:

    a) The driver obtains feedback from control actions and displayed information.

    b) Controls and displays provide visual, auditory and tactile feedback.

The above descriptions of driver perceptions (feedbacks) through the driver interfaces with the vehicle systems are useful for the system designers in developing requirements on the interfaces to ensure proper functioning of the vehicle systems. The ergonomics engineers involved in support the vehicle program need to ensure that all outputs and inputs of the drivers are included in the interface analysis, and systems are designed to reduce driver errors and enhance pleasing perceptions of interfacing with the vehicle.

VEHICLE BRAKE SYSTEM INTERFACES: AN EXAMPLE OF INTERFACE ANALYSIS

An automotive brake (or braking) system illustrated in this section is for a vehicle with front disc brakes, rear drum brakes, ABS capability and a hand parking brake that applies to the rear drum brakes. The braking system was decomposed into four subsystems and major components in each of the subsystems were assumed to be as follows:

1. Hydraulic subsystem:
   a) Brake pedal
   b) Vacuum booster
   c) Vacuum Pump
   d) Master cylinder
   e) Brake fluid reservoir
   f) Brake lines
   g) ABS solenoid valves
2. Mechanical subsystem
   a) Calipers with pistons
   b) Brake pads
   c) Brake rotors (disc)
   d) Wheel hubs
   e) Spindle/axle
3. Parking brake subsystem
   a) Parking handbrake
   b) Parking brake cables
   c) Cams and brake pads
4. ABS subsystem
   a) ABS computer/controller
   b) ABS warning light
   c) Wheel-speed sensors

Other vehicle systems that interface with the braking system are: 1) body system, 2) electrical system, 3) suspension system and 4) powertrain system. The braking system can also be considered as a subsystem of the vehicle safety system. The braking system can also be interfaced with the vehicle exterior lighting system to operate the stop lamps.

The interface diagram of the braking system is presented in Figure 20.5. The interfaces between components of the subsystems and other vehicle systems are shown by arrows and letters placed above or to the right side of each arrow indicate the type of interface. The letter codes used are as follows: P = physical attachment, S = Spatial sharing of space, F = functional interface, E = electrical interface, and M = material transfer (e.g., brake fluid).

Table 20.2 presents an interface matrix of the braking system, its subsystems, components and other interfacing systems. The systems, subsystems and components

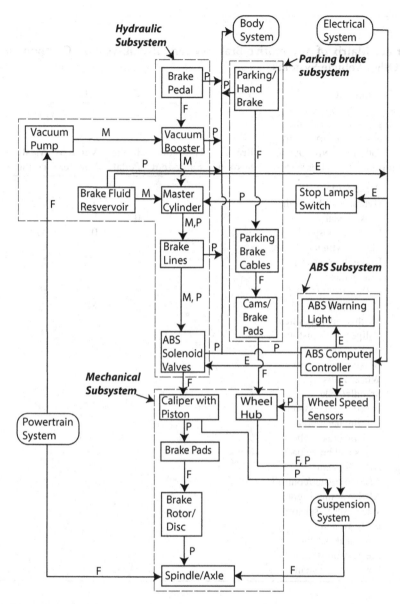

FIGURE 20.5   An interface diagram of an automotive brake system.

are also identified by the following codes: S = system, SS = subsystem, C = component and OS = other system. The letter codes are followed with numbers to identify a system in its first digit and serial number in the second digit. Letter "0" in the interface matrix shows the diagonal with "0" code to designate no interface (same as a blank cell).

TABLE 20.2

Interface Matrix of Automotive Brake System, Its Subsystems, Components and Other Vehicle Systems

Systems, Subsystems and Components: S= System; SSI= Ith subsystem; CIJ=Jth component of ith subsystem; OSI= Ith other system; OCJ= Jth component of other system.		S	SS1	C11	C12	C13
		Brake System	Hydraulic subsystem	Brake Pedal	Vacuum booster	Master cylinder
S	Brake System					
SS1	Hydraulic subsystem					
C11	Brake Pedal			0	F	
C12	Vacuum booster				0	M
C13	Master cylinder					0
C14	Brake fluid resorvior					M
C15	Brake lines					
C16	ABS solenoid valves					P
C17	Vacuum Pump				M	
SS2	Mechanical subsystem					
C21	Calipers with pistons					
C22	Brake pads					
C23	Brake rotors (disc)					
C24	Wheel Hubs					
C25	Spindle/Axle					
SS3	Parking brake system					
C31	Parking hand brake					
C32	Parking brake cables					
C33	Cams and brake pads					
SS4	ABS Ssystem					
C41	ABS computer/controller					
C42	ABS Warning Light					
C43	Wheel speed sensors					
OS1	Body system					
OS2	Electrical system					
OS3	Suspension system					
OS4	Powertrain system					

A quick visual check of the interface matrix shows that most components are sequentially interfaced to the next component (in each row and column) in the first three subsystems. And most of the components are attached to the vehicle body (see column OS1 labelled "Body system").

During the interface analysis the following important issues, trade-off considerations and other observations were made:

C14	C15	C16	C17	SS2	C21	C22	C23
Brake fluid resorvoir	Brake lines	ABS solenoid valves	Vacuum Pump	Mechanical subsystem	Calipers with pistons	Brake pads	Brake rotors (disc)
				F			
	P,M						
0							
	0	P,M					
		0			F		
			0				
					0	P	
					P	0	F
							0

Important Interfaces:

I. Hydraulic Subsystem

   1. The hydraulic subsystem must interface with the powertrain system via the connection of the brake booster to the intake manifold. The powertrain system also includes an electric vacuum pump that will pump up the brake booster if there is insufficient engine vacuum in the manifold to do so. Poor design of this interface may result in the loss of power assisted braking.

TABLE 20.2 (Continued)
Interface Matrix of Automotive Brake System, Its Subsystems, Components and Other Vehicle Systems

Systems, Subsystems and Components: S= System; SSI= Ith subsystem; CIJ=Jth component of ith subsystem; OSI= Ith other system; OCJ= Jth component of other system.		C24 Wheel Hubs	C25 Spindle/ Axle	SS3 Parking brake system	C31 Parking hand brake	C32 Parking brake cables	C33 Cams and brake pads
S	Brake System						
SS1	Hydraulic subsystem			F			
C11	Brake Pedal						
C12	Vacuum booster						
C13	Master cylinder						
C14	Brake fluid resorvior						
C15	Brake lines						
C16	ABS solenoid valves						
C17	Vacuum Pump						
SS2	Mechanical subsystem						
C21	Calipers with pistons						
C22	Brake pads						
C23	Brake rotors (disc)						
C24	Wheel Hubs	0	P				
C25	Spindle/Axle		0				
SS3	Parking brake system						
C31	Parking hand brake				0	F	
C32	Parking brake cables					0	F
C33	Cams and brake pads	F					0
SS4	ABS Ssystem						
C41	ABS computer/controller						
C42	ABS Warning Light						
C43	Wheel speed sensors	P					
OS1	Body system						
OS2	Electrical system						
OS3	Suspension system						
OS4	Powertrain system						

2. The hydraulic system also interfaces with the body system. The pedal box needs to be rigidly mounted to the body. The brake booster also needs to attach in a spot where there is enough room, as it is a fairly large component. If these components are not interfaced with the body correctly, the brake system may not work properly.
3. The hydraulic subsystem also interfaces with the ABS subsystem. If the interface is not done correctly, ABS braking performance may be poor, or complete brake failures may occur.

SS4	C41	C42	C43	OS1	OS2	OS3	OS4
ABS Ssystem	ABS computer/ controller	ABS Warning Light	Wheel speed sensors	Body system	Electrical system	Suspension system	Powertrain system
				P	E	F	
E,F				P	E		F
				P			
				P			
				P			
				P	E		
				P			
						P	
				P	E		
						F	F
				P	E	F	
				P	E		
				F			
				P	E		
0		F	E	P	E		
		0		P	E		
E			0				F
				0			
E		E		P	0	P,F	P
				P,S		0	
				P	E	S	0

II. ABS Subsystem

1. The ABS subsystem interfaces with the electrical system. In most modern cars many other subsystems may react to ABS braking events (transmission shifting, engine power reductions, etc.), and this information needs to be communicated to other electrical modules to ensure the entire vehicle reacts appropriately.

2. The interface with the drivetrain system is necessary to ensure that wheel speed can be appropriately measured at all points. Accurate wheel speed information is necessary to ensure the ABS activates when needed.

3. The interface with the mechanical subsystem is critical to ensure that the required hydraulic pressure is delivered or reduced as needed by the ABS subsystem, and that required braking performance is maintained by the ABS.

III. Mechanical Subsystem

1. The mechanical subsystem interfaces directly with the drivetrain system to decelerate the vehicle. It is important that all components fit together well to ensure that the required braking torque is delivered to the wheels.
2. The interface of the mechanical system to the ABS subsystem is important. The ABS is responsible for delivering the required hydraulic pressure to this subsystem so that the vehicle will decelerate without wheel lock-up on low-friction pavements.
3. The components within this subsystem must interface properly with each other. Improper fit and coordination of components can result in many braking problems, such as rotor warping, premature wear, NVH issues, and so forth.

Design Trade-Offs:

1. An important trade-off is balancing the size of the mechanical subsystem components that interface with the driveline components. Having large calipers, pads, and rotors, which are specially designed to increase braking friction and improve heat reduction are critical to meeting brake performance objectives. The wheel hub, wheel, and suspension all need to be designed to incorporate these components to ensure that brake performance requirements are met. Larger brake system components (i.e., calipers, brake pads and rotors can increase unsprung weight, which can affect vehicle ride and handling performance.
2. The brake pedal and booster need to be rigidly mounted to the vehicle body. The mechanical interface needs to be very robust, i.e., not affected by vibrations, corrosion, temperature changes and high brake pedal actuation forces. This leads to a desire to use a large, heavy brake pedal and linkage to the booster to ensure that the subsystem does not get damaged from heavy usage from aggressive drivers. This leads to a need for large forces in the attachment hardware. Thus, the space required to provide a robust booster must be considered while trading and allocating space for other components in the engine compartment.
3. There is a trade-off between the electrical system cost and ABS pump performance. When active, the ABS pump represents a significant load on the electrical system. As the pump becomes more powerful, the amperage load is greater, requiring larger cables and an alternator to support the load.
4. There is a trade-off between the capacity of vacuum pump required and the cost and the space required to incorporate in the engine compartment. The brake booster relies on engine to provide needed vacuum power assisted

braking. Since engines have been getting more fuel efficient, sometimes the vacuum created is not enough. Thus, an additional vacuum pump may be needed to provide vacuum for the booster in lieu of the engines, especially when operating at higher altitudes. Thus, to provide better braking performance, additional space and electrical load required for the vacuum pump must be considered.

Other Observations:
An observation that can be made from the above example is that many systems and subsystems are often involved in providing basic vehicle functions. Managing the complexity of these systems and interfaces is always a challenge for systems designers and engineers. Large amounts of data are gathered and used in designing all the interfaces for each component in the selected system. Exercises such as development of the interface diagram and matrix can help the component engineers in organizing and understanding the information needed to develop their components and to ensure required functionality in the vehicle. The information gathered would also be useful in developing interface requirements used during the interface design process. An interface requirement must specify the characteristics of the two interfacing entities and how they should perform under a given set of operating conditions.

The driver inputs (DI) and driver outputs (DO) were not shown in the above brake system analysis. The DOs will be driver applied foot-exerted force on the brake pedal and hand-applied force on the parking brake. The DIs will be braking sound (generated in brake drums and by brake pads and brake discs) and the vehicle deceleration felt by the driver.

DESIGN ITERATIONS TO ELIMINATE OR IMPROVE INTERFACES

Reducing number of interfaces will involve a) reducing vehicle features, b) increasing complexity of interfaces by combining two or more interface types (e.g., mechanical attachment also functions as an electrical connector), c) change modality of interface (e.g., change from electrical connection to wireless data transfer).

Improving interfaces requires a lot of brainstorming into areas such as new configurations of systems within the product envelope, breakthrough concepts, applications of new technologies, discarding old designs and carryover entities, reducing weight, size and costs and investments into new interface designs. For example, current trends in driver interface designs involve reconfigurable driver interfaces with new technology displays and controls, more intuitive touch displays, and multifunction controls – which provide the driver options to select his preferred combinations of displays and controls.

SHARING OF COMMON ENTITIES ACROSS VEHICLE LINES

Sharing an entity (system, sub-system or component) across a number of vehicle lines involves standardization (or "commonization") of the shared interfaces so that they can work together (e.g., attach, transmit signals or materials such as fluids or gases) with their corresponding mating entities in a different vehicle. This thus restricts the design (i.e., configuration) of mating entities, which in turn may also affect the performance of the involved systems. For example, use of a common alternator in different vehicle lines would restrict the mechanical and electrical interfacing systems used in different vehicle lines. The commonization will reduce or even eliminate design work associated with designing different alternators and its corresponding connectors which in-turn will reduce design and manufacturing costs; however, it may restrict the overall availability of electric power within the electrical systems of the vehicles that share the same alternator design.

CONCLUDING REMARKS

Interfaces are very important, as they make different entities link and function together to perform the required functions of the product. Interface design and production of interfaces involves expenditures of time, money and specialized resources. It is important to remember the following point that wiring harness producers often make: "The major costs increases are not in the increase in length of the wiring harness, but they are in the complexity of the 'connectors' at both the ends (i.e., the interfaces) of the harness."

REFERENCES

Bhise, V. D. 2023. *Designing Complex Products with Systems Engineering Processes and Techniques.* Second Edition. ISBN: 978-1-032-20369-0. Boca Raton, FL: CRC Press.

Lalli, V. R., R. E. Kastner and H. N. Hartt. 1997. Training Manual for Elements of Interface Definition and Control. NASA Reference Publication no.1730.

National Aeronautics and Space Administration. 2007. *NASA Systems Engineering Handbook.* Report no. NASA/SP-2007-6105 Rev1. NASA Headquarters, Washington, D.C. 20546. http://ntrs.nasa.gov/archive/nasa/casi.ntrs.nasa.gov/20080008301_2008008500.pdf (accessed October 15, 2012).

Pimmler, T. U. and S. D. Eppinger. 1994. Integration Analysis of Product Decompositions. In *Proc. ASME 6th Int. Conf. on Design Theory and Methodology*, Minneapolis, MN.

Sacka, M. L. 2008. A Systems Engineering Approach to Improving Vehicle NVH Attribute Management. Master's Thesis for M. S. degree in Engineering Management at the Massachusetts Institute of Technology, Cambridge, MA.

21 Detailed Engineering Design During Automotive Product Development

INTRODUCTION

Many steps need to be performed in the detailed engineering design (DED) phase of an automotive product development program. Many of these steps can be performed concurrently, involving personnel from different disciplines and departments. The sequence of steps and outputs of the steps are generally identified in the systems engineering management plan, product development process, and/or engineering design manuals prepared by the company developing the product. The ergonomics engineers assigned to the vehicle program follow the progress of teams involved in designing various vehicle systems and provide necessary ergonomics data required to resolve the issues. The detailed engineering design process generally begins after the concept selection phase. However, some early engineering design work is generally performed during concept development to ensure that the product concept will be technically and economically feasible, and its components could be manufactured and assembled using equipment available within the company or its supplier facilities.

Completion of the DED phase means all components within all systems of the complex product are designed (or developed) which include a) engineering drawings (usually using a CAD software showing different views and dimensions of a solid model for each component or a system, b) CAE analyses and evaluations – for example, structural analysis, thermal analysis, aerodynamic/fluid flow analysis, electrical and electronic circuits analyses – and c) material specifications (i.e., material or composition of metals, plastics, rubbers, etc.) to be used to produce each component including any special treatments such as heat treatment and surface coating, d) applications of principles related to design for manufacture, design for assembly, and design for sustainability, and e) evaluations and verification tests – usually engineering tests using computer models and/or physical tests – conducted to ensure that each component meets its stated functional, manufacturing, assembly, safety and cost requirements (i.e., all attribute requirements cascaded down from the product level). The developed product thus conforms to the "balanced product design" shown in the systems engineering process (see Volume 1, Chapter 2, Figure 2.2).

DOI: 10.1201/9781003485605-8

ENGINEERING DESIGN

Engineering design requires integration (i.e., simultaneous consideration) of many issues such as, a) meeting many requirements – based on customer needs, governments regulations, and company/industry standards – and b) involving multiple disciplines to ensure consideration of issues and applications of analysis techniques from different engineering and related disciplines such as ergonomics, electrical, mechanical, materials, software, chemical, safety, manufacturing, assembly, and sustainability, c) safety analyses and evaluations to reduce or eliminate possible accidents and injuries that can be caused by the product and the possibilities of product liability cases during its use, d) conducting many design iterations to consider many product attribute requirements and trade-offs between different attributes in various combinations of configurations (called design synthesis in the SE process, see Volume 1, Figure 2.3), and f) a lot of engineering analyses, design reviews, and testing (subjective and objective) to verify that the components within all lower-level systems and the whole product will meet their stated attribute requirements.

Customer needs, attributes, attribute requirements (product level), concept design, system requirements, systems design, sub-system requirements, and design of lower-level systems down to components are all considered during the design activities of team members involved in engineering design. Their output thus is a fully functional product design that meets all the requirements.

Many of the entities such as vehicle systems, sub-systems and/or their components are generally designed and produced by different suppliers. Their personnel are integrated into the design team and, hence, they participate in design team meetings by sharing and exchanging information to ensure their designs can be packaged into all affected and interfaced vehicle systems.

There is no unique process to accomplish this complex engineering design phase. But the process typically involves the following steps: Note: The order and number of steps can vary between organizations, products and amount of concurrent engineering work done among various design teams.

1. Documenting and communicating the specifications of the product to be developed during the early phases (e.g., advanced design, market research and concept design) to the team members, refining the specifications and ensuring that those collectively define the product to be engineered. The product specifications include its characteristics, such as dimensions, functionality, performance, and features, which cover its all attribute requirements. These tasks are based on:

    a) Collecting customer needs data from marketing research and benchmarking data from engineering analysis of competitors' products
    b) Searching for all government requirements (federal, state and local) applicable to the product during its entire life cycle
    c) Compiling and reviewing of customer feedback and warranty data on existing products available in the market to meet the same customer needs

2. Conducting functional analysis and requirements analysis for the product
3. Development and documentation of product attributes and attribute requirements

4. Creating alternate product design configurations, including CAD drawings and models

5. Refining product configuration and creating decomposition trees showing systems and lower-level systems down to component level

6. Risks and trade-offs associated with each vehicle concept are identified, and acceptable design configuration is selected by involving all affected disciplines in design review meetings.

7. Cascading requirements from the product level to all lower-level systems down to components

8. Identifying all interfaces (between systems and their lower-level systems and components) and developing requirements on all interfaces (see Chapter 20)

9. Using the systems engineering "V" model to ensure that design of systems, sub-systems and components are developed using the top-down approach. Any carryover components to be used should be included in the analyses.

10. Conducting detailed engineering analyses involving material selection, evaluation of functional capabilities using available models and CAE techniques (e.g., structural analysis, applying design for manufacture and assembly techniques and guidelines for design for sustainability), and building prototype parts and testing to verify that the design of each component meets its stated requirements. CAE analyses can involve evaluation of designs of mechanical structures including CAD modeling and analyses – for example, stress testing and structural simulations, aerodynamics and thermodynamic analyses, thermo-mechanical analyses, fluid flows (computational fluid dynamics), fatigue and durability analyses, acoustics/sound and noise, vibrations and harshness analyses, electrical and electronic circuits, control systems analyses, dynamic simulations and analysis of moving components – biomechanical and ergonomics analyses, toxic exposures, life cycle costing, and so forth.

11. Conducting Failure Modes and Effects Analyses (FMEA) before the design is released for production. The results of the FMEAs are reviewed by the management and changes are made to the designs to meet the accepted level of risk priority number.

12. Assembling verified components into lower-level systems and lower-level systems into higher-level systems, and tests are performed to ensure that all the assembled sub-systems and systems meet can fit within the product space without any interferences and meet required clearances while meeting their respective functional requirements. Results of the tests are reviewed by engineering management personnel.

13. Finally, the assembled systems are further assembled to create the whole product.

14. A number of design reviews are conducted during various stages of the above design steps.

15. The whole product is further tested to verify that it meets all stated requirements and validated to ensure that the customers like and are satisfied with the product before it is released for production.

The above steps can be better understood by studying engineering design examples of actual complex products. Thus, the following section provides descriptions of steps applied to six different complex products.

SIX PRODUCTS EXAMPLE

The tasks that are performed during the detailed engineering phase depend upon the product, its characteristics and manufacturing, construction and assembly processes. Six different complex products are presented here to illustrate their design related issues and similarities in engineering design processes.

The following six complex products are included in Table 21.1. The major systems of these products are listed below.

1. *Washing Machine*: Body system, clothes holding tub and drive system, water and detergent management system, washing cycle control system, controls and displays system
2. *Refrigerator*: Body system, exterior panel and trim system, insulation, interior trim and food storage system, refrigeration system, electrical power and control system, water management and ice making system, and interior lighting system
3. *Laptop Computer*: Chassis system, microprocessor system, data storage and retrieval system, data input system (keyboard, mouse and microphone), data output system (display and audio system), wireless communication system, power system, and cooling system
4. *Automotive Product*: Body system, chassis, steering and suspension system, powertrain system, cooling system, braking system, electrical system, fuel system, exhaust system, climate control system, airflow management system, driver interface system, seating system, lighting system, safety system, and luggage/cargo storage system
5. *Wind Turbine*: Foundation system, tower structure system, nacelle system, wind turbine (blades and rotor) system, gear box system, generator system, step-up transformer system, air flow direction and wind velocity sensing system, turbine control system, electrical system, and safety systems including steps, tethering, handholds and ladders
6. *Natural Gas Fired Power Plant*: Natural gas storage system, gas and air input and control system, plant housing system, gas turbine system, power generator system, power plant control system, plant lighting system, piping systems, heat exchanger, secondary steam turbine, and generator system, power distribution system, water management system, auxiliary power system, and plant safety/security system

Table 21.1 also provides some important systems engineering considerations of the products such as variables used to measure their output capacity, special features, product attributes, disciplines involved during design phases, engineering analyses

TABLE 21.1
Characteristics of Six Complex Products and Disciplines and Analyses Involved in Their Detailed Engineering

Complex Product	Washing Machine	Refrigerator	Laptop Computer	Automotive Product	Wind Turbine	Natural Gas Fired Power Plant
Systems within the product	Body system, clothes holding tub and drive system, water and detergent management system, washing cycle control system, controls and displays system	Body system, exterior panel system, interior trim and food storage system, refrigeration system, electrical power and control system, water management system, water and ice making system, and interior lighting system	Chassis system, microprocessor system, data storage and retrieval system, data input system, data output system, wireless communication system, power system and cooling system	Body system, chassis, steering and suspension system, powertrain system, braking system, electrical system, fuel system, cooling system, exhaust system, climate control system, driver interface system, lighting system, safety system and luggage/cargo storage system	Foundation system, tower structure system, nacelle system, wind turbine (blades and rotor) system, gear box system, generator system, step-up transformer system, air flow direction and wind velocity sensing system, turbine control system, electrical system, and safety system including steps, handholds and ladders.	Natural gas storage system, gas and air input and control system, gas turbine system, power generator system, power plant control system, plant lighting system, heat exchangers, secondary steam turbine and generator system (for combined cycle plant), power distribution system and plant safety/security system

(continued)

TABLE 21.1 (Continued)
Characteristics of Six Complex Products and Disciplines and Analyses Involved in Their Detailed Engineering

Complex Product	Washing Machine	Refrigerator	Laptop Computer	Automotive Product	Wind Turbine	Natural Gas Fired Power Plant
Capacity	Weight or volume of wash load (3.5 to 5.0 cf)	Volume of refrigerated (6–30 cf) and freezer (1–6 cf) compartments	Processor speed (up to about 3.6 GHz) and data handing capacity), memory capacity (128 GB–1TB SSD, display size (13–16")	Number of passengers (4–7), volume and weight of load carrying capacity, towing load	1500–3500 kW of electricity produced per hour per turbine	300–1000 MW of electricity produced per hour
Special Features	Number of settings for fabric types, temperatures, wash cycle time	Adjustable shelves, lighting, auto defrost, garageability	Memory type and storage, processor type and speed, keyboard lighting, touch screen, weight	Interior trim (e.g., leather seats), electronic features, LED lighting, reconfigurable interior and cargo compartments, etc.	Blade defrosters, auto shut-off, elevator in shaft, remote controls, etc.	Emission controls (carbon capture and sequestration), combined cycle, automated control system, etc.
Some Attributes of the Product	Washing Capacity, Packaging Space, Ergonomics, Washing efficiency, features (e.g., settings, displays,	Cooling and Freezer Space, Exterior size, Energy Usage, Features (shelves, adjustability, lighting, Noise and Vibrations, Cost/Price	Weight, Display Size, Overall Size, Processor Capacity, Memory Capacity, Sound Quality and Loudness, Noise, Special Features (e.g., lighted keys, 2 in 1 hinges)	Overall Shape/Body Style, Size, Styling, Performance, Vehicle Dynamics, Aerodynamics, Fuel Economy and Emissions,	Electricity generation capacity (1.5–3.5 MW), Capacity Factor(28–35%), Exterior and Interior Package, Noise and Vibrations, Safety, Cost/Price	Electricity generation capacity (MW), Capacity Factor(60–87%), Exterior and Interior Package, Noise and Vibrations, Emissions, Safety, Cost/Price

Disciplines Included in Engineering Design Work	top vs. side loading), Noise and Vibrations, Cost/Price	Mechanical Engineering, Electrical Engineering, Manufacturing and Assembly, Chemical Engineering, Marketing and Finance	Mechanical Engineering, Electrical Engineering, Manufacturing and Assembly, Chemical Engineering, Marketing and Finance	Mechanical Engineering, Electrical Engineering, Manufacturing and Assembly, Chemical Engineering, Marketing and Finance	Costs/Price, Noise, Vibrations and Harshness, Safety, Customer Lifecycle, Product and Process Compatibility	Mechanical Engineering, Materials Engineering, Electrical/ Electronics Engineering, Manufacturing and Assembly, Chemical Engineering, Safety Engineering, Marketing and Finance	Mechanical Engineering, Civil Engineering, Aerodynamics engineering, Manufacturing and Assembly, Chemical Engineering, Safety Engineering, Marketing and Finance	Mechanical Engineering, Civil Engineering, Electrical Engineering, Manufacturing and Assembly, Chemical Engineering, Safety Engineering, Marketing and Finance

(continued)

TABLE 21.1 (Continued)
Characteristics of Six Complex Products and Disciplines and Analyses Involved in Their Detailed Engineering

Complex Product	Washing Machine	Refrigerator	Laptop Computer	Automotive Product	Wind Turbine	Natural Gas Fired Power Plant
Specific Examples of Engineering Analyses Performed during Detailed Engineering Design Phase	Packaging and CAD Modeling, Heat and Fluid Flow Analyses, Body Structure and NVH, DFM, DFA, FMEA, LCA, LCCA, Verification and Validation Testing	Packaging and CAD Modeling, Heat and Fluid Flow Analyses, Body Structure and NVH, DFM, DFA, FMEA, LCA, LCCA, Verification and Validation Testing	Electrical and Electronics Architecture, Packaging and CAD Modeling, Heat and Fluid Flow Analyses, Body Structure and NVH, DFM, DFA, FMEA, LCA, LCCA, Verification and Validation Testing	Packaging and CAD Modeling, Heat and Fluid Flow Analyses, Body and Chassis Structure Design and NVH, DFM, DFA, Vehicle Dynamics, FMEA, LCA, LCCA, Verification and Validation Testing, Safety Evaluations	Packaging and CAD Modeling, Heat and Fluid Flow Analyses, Civil Engineering, Structural and NVH, Electrical Systems Architecture and Design, DFM, DFA, FMEA, LCA, LCCA, Verification and Validation Testing	Civil Engineering, Plant Design, Packaging and CAD Modeling, Heat and Fluid Flow Analyses, Chemical Engineering, Body Structure and NVH, DFM, DFA, FMEA, LCA, LCCA, Verification and Validation Testing
Key Validation Tests	Cleanliness of Clothes, Noise level	Temperature Maintenance, Power Consumption	Processing Speed, Uploading/ Downloading Speed, Display Readability, Keyboard Comfort	Fun-to-drive, Interior Space, Fuel Consumption, Safety Ratings, Exterior Styling, Ride and Handling	Power Generated, Noise Level, Capacity Factor	Power generated, Capacity Factor, Fuel Consumption
Product Life Cycle (years) (Approximate)	8–15	10–20	3–8	8–14	20–30	25–35
Price ($) (Approximate)	$350–$1200	$500–$3000	$450–$1500	$25,000–$65,000	$1200 (on-shore) - $4300 (off-shore)/ kW	$1000–$2500/kW

methods used during design, key validation tests to be conducted to determine if the customers will be satisfied with the products, product life cycle and price.

The customer needs should drive the design of each of the above products. The customer needs can be better understood from the required level of product capacity, special features, and product attributes (see Table 21.1). The overall technical workload during the concept design and detailed engineering phases of these products will depend upon technologies and complexity involved within the product. The complexity of the product will depend upon the total number of entities (i.e., systems, subsystems and lower-level systems down to the component level) within the product and number of interfaces between the entities. In addition, optional features demanded by the customers will include other enhancements added to the basic product design. The capacity (see second row of Table 21.1) of the product will be estimated from the customer demand. The overall volume of its envelope can be estimated from customer demand (rate of the product output) and technology used to operate the product.

After the data gathering on customer needs, benchmarking, and documentation of the product specifications, the overall product concept is created. Packaging space needed for creation of the product is usually the first consideration in engineering design where the design engineers along with other team members decide on the shape and size of space required, that is, the overall envelope of the product. The envelope is created by understanding other existing products available in the market segment by benchmarking and detailed functional analysis. Technological trends and latest developments are all important considerations to determine improvements in the functionality of the product, its overall configuration, and their effects on other product attributes (e.g., weight, costs, performance).

The engineers make a number of design related decisions by answering questions such as:

a) How would the systems within the product be configured to fit within the space (envelope)?
b) Would the customers and users of the product enjoy using the product?
c) Can all the systems with their respective subsystems be configured and interfaced within the allocated spaces for each subsystem?
d) Are sufficient clearance spaces provided between systems for the maintenance personnel to service the systems?
e) Would the emissions during processing of materials and use of the product create harmful effects?
f) Would the cost of the product be within the budgeted amount?

As the design of each of the subsystems progresses simultaneously within different teams, the teams meet frequently to ensure that all the subsystems can fit and function within the allocated spaces. Design reviews and verification tests (CAE analyses, simulations, creation of prototype parts and physical testing of the parts and their assemblies) are conducted to ensure that the components and assembled systems (lower to higher level) meet their respective attribute requirements (see Figure 21.1)

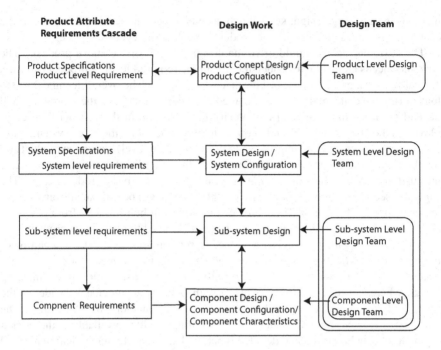

FIGURE 21.1   Product specifications to component design.

The left side of Figure 21.1 shows that product specifications are developed first to ensure that the product is designed to meet its specifications. Next the specifications of the systems within the product are developed by cascading the product level requirements into the system's level requirements. Similarly, the systems level requirements are cascaded down to lower-level systems (sub-systems and components). The middle column shows the design work, and the third column shows the teams involved in design. The teams use CAD to illustrate shape and size (dimensions) of the components and determine materials for the components. A number of CAE analyses and simulation tests (e.g., stress, strength and durability evaluations) along with design for manufacturing, design of assembly, and design for sustainability evaluations are conducted to ensure that the components can be assembled into lower-level systems, the lower-level systems can be assembled into the product and the product will function to meet its specifications. (Note that these steps are incorporated in the left side of the systems engineering "V" model.)

SYSTEMS ENGINEERING "V" MODEL

The systems engineering "V" model presents all important steps in the life cycle of a product or a system. The model is presented in Volume 1, Chapter 2 and illustrated in Figure 2.5. The model is known as the Systems Engineering "V" Model because the steps are arranged in a "V" shape with succeeding steps shown below or above the preceding steps and staggered in time (see Bhise (2017, 2023) and Blanchard and

Fabrycky (2011) for more details). The model in Volume 1, Chapter 2 is described for the development of an automotive product.

The systems engineering "V" model thus shows the locations of the "Design and Engineering" and "Verification, Manufacturing and Assembly" tasks on the time axis along as the left and right sides of the "V". The "Operation and Refinement" and "Retirement and Disposal" tasks are shown on the right side of the "V". The model also shows how the requirements developed during the design process are used for verification of the components, sub-systems, systems, and the whole system, as they are assembled (see horizontal arrows from left side of the V to the right side of the V). The design process uses the "top-down" approach, that is, it begins with the development of the product concept on the top left and ends at the top of the "V" on the right side with the assembled and product ready to be shipped to the automotive dealers.

ACTIVITIES IN ENGINEERING DESIGN

Figure 21.2 illustrates the major steps in detailed engineering design of a component of an automotive product. The top of the diagram in Figure 21.2 shows that the process of designing a product always begins with creating the overall product

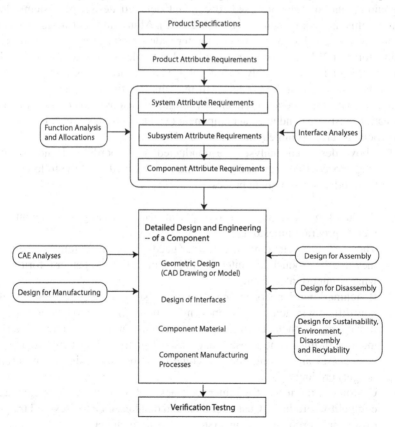

FIGURE 21.2   Component design process.

specifications. The product specifications should be documented and provided to all the team members to ensure that each member has a clear understanding of details about the product, that is, what the product should accomplish. For example, if the product is a midsize SUV for the US market the product specifications should provide details such as its overall exterior dimensions, weight, important interior dimensions, powertrain and transmission capabilities, suspension type and important features. The vehicle attribute managers use the vehicle specifications and develop detailed requirements on all attributes and sub-attributes of the vehicle. The vehicle attribute requirements are cascaded down to vehicle systems (e.g., body system, powertrain system, chassis system, climate control system, and electrical system) and their lower-level systems such as subsystems and sub-sub-systems down to the component level. As the vehicle systems are decomposed into lower-level systems detailed function analysis is performed to determine how various functions of the vehicle should be allocated (or divided) between vehicle systems. As the configuration of the systems decomposition is being developed, interface analyses (e.g., interface diagrams and interface matrices) are developed to understand interfaces between different levels of systems and components to specify interface requirements.

All the above details about systems decomposition and information generated in requirements analyses are used to conduct detailed design of systems, lower-level systems and components (see Figure 21.2). Many different design tools such as computer aided design (CAD), computer aided engineering (CAE), design of manufacturing (DFM), design for assembly (DFA), design for sustainability (DFS, including design for disassembly and recycling) are applied concurrently by different specialists to ensure that the product will meet its functional requirements and sustainability goals (e.g., reductions in energy usage, emissions, and costs). A number of verifications tests are conducted at component level and progressively at higher levels as components are assembled into higher level systems.

The above described analyses are conducted in a coordinated and concurrent engineering process that is guided by information obtained from the following activities in the product development process:

1. Product Program Plan (tasks to be performed and timings of different events, that is, program milestones).
2. Thorough understanding of customer needs: Visiting customers, seeing how they use the product, and interviewing the customers to gather information on what is important to them.
3. Benchmarking: Understanding studying existing product designs in terms of how different systems and components in these existing products are designed, for example, their configuration, materials, unique features, interfaces, manufacturing methods used, and costs. Understanding their strengths and weaknesses and setting goals to come up improved design of the proposes (target) product.
4. Customer experience: Customer feedback and warranty rates on existing and competitive products: Customer likes and dislikes, their causes, and frequency of warranty problems are understood by team members.

5. Design trends: Trends in design, technologies and government regulations are studied.

6. Program management tasks: These tasks involve understanding and implementation of important design considerations to guide and control activities related to product configuration, functionality, performance features, and program timings and budgeting activities.

7. Product Attribute management: The attribute engineers are responsible for ensuring that attribute requirements are met during the design activities (Note that ergonomics is one of the attributes of the vehicle).

8. Systems engineering management plan (SEMP): The systems engineers develop the SEMP in early phases, and it is followed to manage design team activities and program management functions.

9. Coordination of personnel and teams: systems engineers and program management ensures that team members from required disciplines/departments are represented in multi-functional teams, teamwork is managed to follow the SEMP, management of teams and communications between teams (e.g., meetings) is facilitated, manufacturing and assembly personnel are available during the design process, data sources, and suppliers are integrated into the design process.

10. Activities of the design teams: The activities involve the following:

    a) Design teams generally create system decomposition trees to understand all sub-systems and lower-level systems down to the components (see Volume 1, Figure 2.1 in Chapter 2 for an example of the decomposition tree).

    b) QFD charts are created for major systems, subsystems and components to understand customer needs and cascading the customer needs into engineering specifications. The QFD tool also provides targets for engineering specifications and identified most important functional specifications. Chapter 3 in Volume 1 provides more details on the QFD.

    c) Interface diagram and interface matrix are also developed to understand interfaces between various systems, sub-systems and components. The requirements for interfaces are developed and trade-offs that need to be considered during designs are identified. Chapter 20 provides more details on the interface analyses.

    d) CAD models are developed by package engineers with the help of many specialized engineers. The CAD models and product data are controlled to ensure that unauthorized persons do not make changes in the designs and the design team members get the latest verified design data. Design changes are communicated to all teams and available within the program management systems.

    e) Many design tools such as design for manufacturing (DFM), design for assembly (DFA), design for sustainability (DFS), and failure modes and effects analysis (FMEA) are used to study and improve design details as all lower-level and upper level systems are refined.

11. Simultaneous (concurrent) design and engineering approach and iterative design (need some iterations but not too many to slow down the progress) steps are incorporated.

12. Attribute managers ensure that team members understand engineering considerations and requirements including test procedures.

13. Design reviews are conducted in a timely manner to meet product functionality and program timings.

14. Evaluations are conducted to ensure that the designed product will meet customer needs by performing required design reviews, and tests (computer and physical).

CONCLUDING REMARKS

The overall phase of detailed engineering of a complex product is very challenging. It involves consideration of all engineering and manufacturing requirements concurrently with the help of inputs from many multidisciplinary teams to ensure that all product requirements, trade-offs between product attributes and risks are studied, and the configuration of the whole product along with the design of all systems within the product are developed. The complexity of the tasks involved in this phase increases with the increase in the number of systems, lower-level sub-systems down to every component in each system within the product and interfaces between the many systems and components. The complex problem is solved by conducting many design iterations with inputs from many engineering analyses, design reviews and tests to ensure that the design will meet all stated design requirements. It is therefore very important that the personnel working in this phase (including ergonomics engineers) understand the overall process, their tasks and available problem solving techniques by constantly communicating, collaborating, and seeking advice from experts from various engineering specialties, including systems engineering, which coordinates the activities of different teams through the systems engineering management plan.

REFERENCES

Bhise, V. D. 2017. *Automotive Product Development: A Systems Engineering Implementation.* ISBN: 978-1-4987-0681-0. Boca Raton, FL: CRC Press.

Bhise, V. D. 2023. *Designing Complex Products with Systems Engineering Processes and Techniques.* Second Edition. ISBN: 978-1-032-20369-0. Boca Raton, FL: CRC Press.

Blanchard, B. S. and W. J. Fabrycky. 2011. *Systems Engineering and Analysis.* Upper Saddle River, NJ: Prentice Hall PTR.

22 Vehicle Evaluation, Verification and Validation

OBJECTIVES AND INTRODUCTION

The objective of this chapter is to provide the reader an understanding of what is meant by product evaluation, how evaluations are conducted to verify that the product will meet its stated requirements, and to validate that the right product is developed. The chapter also presents Information on types of evaluations and various evaluation issues and covers a number of ergonomic evaluation methods used in the verification and validation processes of automotive products.

WHY EVALUATE, VERIFY AND VALIDATE?

It is important that we design the right products that will meet customer needs, and customers will like them. The right products also build customer loyalty and thus customers will continue to use products made by the manufacturer and purchase the next products made by the manufacturer. The product development team must have a clear focus on meeting the stated customer needs. Otherwise, the designed product may deviate from its goal; for example, the product may become too difficult to use, too heavy, too bulky, consume too much energy, produce much more pollutants as compared to other benchmarked products, be too costly, and too boring (not fun to use). During the product development process, we must continuously check and ask the questions if the product possesses the right characteristics and whether the product being designed is indeed the right product for the customers.

TESTING, VERIFICATION AND VALIDATION

Testing is an activity undertaken by using well-established procedures to obtain detailed measurements and data on the characteristics and performance of the product, its systems, subsystems, or components. The collected test data are analyzed to determine if the product, its systems, subsystems, or components meet their stated requirements (i.e., specified during the design process). Thus, a test can be conducted at the product, system, subsystem, or component levels to determine if one or more of the requirements at its corresponding level are met. A test can be performed by using computer models, simulations, prototypes, or physical working samples of the

DOI: 10.1201/9781003485605-9

hardware representing the product, its systems, subsystems, or components. Testing methods can be used for verification or validation purposes.

Verification is the process of confirming that the product, its systems, and its components meet their respective requirements. The aim of the verification is to ensure that the tested item (product, its system, subsystem, or component) is built right, that is, it meets its requirements.

Validation, on the contrary, is the process of determining whether the product functions and possesses the characteristics as expected by its customers when used in its intended environment. The aim of the validation process is to ensure that the right product is designed, and the product can be used and liked by its intended customers.

DISTINCTIONS BETWEEN PRODUCT VERIFICATION AND PRODUCT VALIDATION

From a process perspective, the product verification and product validation processes may be similar in nature, but the objectives are fundamentally different. Verification of a product shows proof of compliance with its requirements – that the product can meet each "shall" statement as proven through performance of a test, analysis, inspection, or demonstration. Validation of a product shows that the product accomplishes the intended purpose in the intended environments – that it meets the expectations of the customer and other stakeholders as shown through performance of a test, analysis, inspection, or demonstration.

Verification testing relates back to the approved requirements (which generally include all attribute requirements; see Volume 1, Chapter 2) set in the early phases of product development. The verification testing can be performed at different stages in the product life cycle. The approved specifications, drawings, computer-aided design (CAD) models, physical bucks (or mock-ups), prototypes representing the designs of systems, subsystems, and components and their configuration documentation establishing the baseline of the product which may have to be modified at a later time. Without a verified baseline and appropriate configuration controls, later modifications could be costly or cause major performance problems.

Validation testing is conducted under realistic conditions (or simulated to represent the real usage conditions) for the purpose of determining the effectiveness and suitability of the product for use in the mission operations by typical users. The selection of the verification or validation methods are generally based on the engineering, program management, purchasing, or the certifying agency's judgment as to which is the most effective way to reliably show the product's conformance to requirements or that it will operate as intended.

OVERVIEW ON EVALUATION ISSUES

A product such as an automobile is used by a number of users in a number of different usages. To ensure that the product being designed will meet the needs of its customers, the engineers must conduct evaluations of all product features under all possible usages. Usage can be defined in terms of each task that needs to be performed by the user to meet a certain objective. A task may have many steps or subtasks. For example, the task of getting into a vehicle would involve a user to perform a series of subtasks such as (1) unlocking the door; (2) opening the door; (3) entering the vehicle and sitting in the driver's seat; and (4) closing the door. The evaluations are conducted for a number of purposes such as (a) to determine if the product meets its perform- ance requirements; (b) to determine if the users will be able to use the product and its features; (c) to determine if the product has any unacceptable features that will gen- erate customer complaints after its introduction; (d) to compare the user preferences for a product or its features with other similar or benchmarked products; and (e) to determine if the product will be perceived by the users to be the best in the industry. The purpose of this chapter is to review methods that are useful in the evaluations of products.

The evaluations can be conducted by collecting data in a number of situations. Figure 22.1 presents a simplistic flow diagram illustrating how the customers are involved in the verification and validation processes. Box no. 1 (top left in Figure 22.1) shows that the process begins with understanding the needs of the customers. The list of customer needs is translated into a list of product attributes (Box no. 2). To ensure that the product possesses the attributes, requirements are developed for each product attribute (Box no. 3). The requirements are used during the product develop- ment process to create product concepts and product designs; and various systems, subsystems, and components are tested to verify that they meet their respective requirements (Box no. 4). When the entire assembled product is available (usually as an early prototype), it is tested by using a number of customers and experts to validate the product (Box no. 5). After the product is sold to the customers, their feedback (through customer clinics, complaints, warranty and repair data, usage experience through dealers, and accident data) are gathered (Box no. 6). The customer feedback

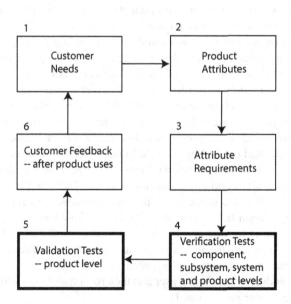

FIGURE 22.1   Verification and validation tests and the customers.

data are used in developing (or refining) the list of customer needs (Box no. 1) for incorporating changes in the future models or versions of the product.

TYPES OF EVALUATIONS

Some examples of types of product evaluations conducted for product verification and validations are given in the following:

1. *Functional evaluations:* These evaluations consist of tests on individual components, subassemblies, and assemblies to ensure that they would perform the required functions or tasks under selected environmental conditions and meet their respective performance specifications.
2. *Durability tests:* The assembled subsystems, systems, and products are tested under the most demanding actual situations (e.g., at minimum and maximum operating temperatures, high workloads, high speeds, and maximum electromagnetic fields) over a large number of work cycles (maximum expected work cycles during the life cycle, for example, 10 years or 150,000 miles for the passenger cars. Note: Many tests may be conducted under accelerated-life-testing conditions).
3. *Assembly evaluations:* Assembly tasks are performed and evaluated under actual work situations or simulated situations (e.g., using 3-D CAD models of the assembly workstations, with the product, tools, and operators/robots) to check that the components can be properly assembled into systems and the systems can be assembled into products with a minimum number of movements and reorientations of the entities and hand–body motions of assemblers (or robots). Special attention should be paid to many steps in the tasks such as, a) getting the components to the workstations, b) transporting or moving the components to the assembly fixtures, c) orienting and clamping the components in the fixtures and d) joining the components using fasteners, welds, and glues to ensure that sufficient clearance spaces are available for access for easy movements of the components, tools, and arms of a human assembler or a robot. The timings of different events and time required for assembling, inspecting for proper assembly, and so forth, are also checked.
4. *Tests involving human subjects:* Market researchers and human factors engineers conduct tests using representative customers or users. Some products can be operated or used by a single human operator, whereas other complex products such as commercial airplanes and ships require crews where coordination of actions and communications among individuals are important for successful operation of the missions. Some examples of simple test situations involving human subjects are described in the following:
    a. A product (or one or more of its systems, chunks, or features) is shown to a user and the user's responses (e.g., facial expressions and verbal comments) are noted (or recorded).
    b. A product is shown to a user and then responses to questions asked by an interviewer are recorded.

    c.   A customer is asked to use a product and then to respond to a number of questions included in a questionnaire or are asked by an interviewer and recorded.

    d.   A user is asked to use a number of products, and the user's performance in completing a set of tasks on each of the products is measured.

    e.   A user is asked to use a number of products and then to rate the products based on a number of criteria (e.g., preference, usability, accommodation, effort, and comfort).

    f.   A sample of customers are provided with instrumented products that record customer behavior and product outputs: for example, video recording of driver behavior and customers' performance as the participants drive where they wish, as they wish, for weeks or months each. This is probably the only valid method to discover what drivers/customers actually do over time in the real world.

5.   *Evaluations after product sales:* Tracking how the product is perceived by its customers based on their product usage experience is a very important approach in validating the product. Manufacturers generally keep track of customer feedback and on repair and warranty data to determine customer satisfaction. The best measure of validation is probably the number of sales in periods after the product introduction. Continued higher sales volumes indicate that customers liked the product and continue to purchase the products. Repurchases of the same product after the original product has been out of service due to reasons such as end of life (or excessive usage or wear), losses in accidents, or simply additional purchases of the same product for expanded customer base are strong indications of customer acceptance and loyalty.

The aforementioned examples illustrate that an engineer can evaluate a product or its features by using a number of data collection methods and measurements. The evaluations can be conducted by using computer models, simulations, physical models, early prototypes, production prototypes, and the products during and after use. The environment for the evaluations can be simulated in laboratory or field tests can be conducted at selected locations to meet the required environmental conditions (e.g., at different combinations of light, sound, vibrations, static/dynamic, temperature, humidity, and snow/rain levels).

EVALUATIONS DURING PRODUCT DEVELOPMENT

During the entire product development process, a number of evaluations are conducted to ensure that the product being designed will meet the needs of the customers. The attribute requirements and design issues need to be systematically studied to ensure that all design requirements are considered, and appropriate evaluation methods are used. The results of the evaluations are generally reviewed in the product development process at different milestones with various design and management teams.

Figure 22.2 illustrates the timings of the product verification and validations tests (shown by circles) in the systems engineering "V" model of the product development

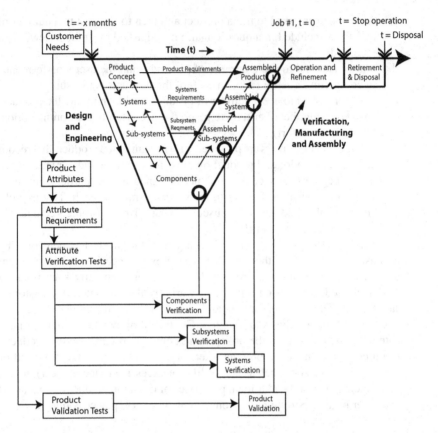

FIGURE 22.2   Verification and validations test during the product development program.

process. The figure shows that customer needs are documented before the program initiation (at $-x$ months). Product attributes are developed from customer needs. Attribute requirements (developed from customer needs) are used to create attribute testing and product validation plans. Attribute verification tests are conducted to verify the components, subsystems, and systems. Product validation tests are conducted on fully assembled products available as early prototypes or production units. The verifications tests are generally engineering tests using simulations, laboratory, or field tests employing test software, test equipment, and test operators. The outputs of the tests (measurements) are compared with the minimum and maximum acceptance criteria values specified in the attribute requirements. The validation tests are usually conducted by using customers, users, and experts to ensure that the product meets all the attribute requirements. Appropriate statistical tests are also conducted to assure that the data collected in the tests meet the requirements.

VERIFICATION PLAN AND TESTS

A verification plan is needed to demonstrate that all product requirements are evaluated, and the targets are met. Product requirements are generally derived from the product specifications and attribute requirements. These requirements can also be selected from available internal (e.g., corporate) manuals of engineering specifications, standards, and brand requirements. The responsibility of documentation of verification plans for various entities is generally assigned to the design release engineers and the attribute engineers.

The planning of the verification testing involves identification of the appropriate methods and test equipment that are needed to verify each requirement. The verification plan must also include the number of product samples that need to be tested to verify each requirement. The verification plan also includes other data such as recommend test dates, specific requirement(s) being verified, the target or acceptance criteria, and the required statistical tests to be performed to support the results of the tests. The engineering managers are usually responsible for approval of the verification test methods and the test plan.

The verification plan also identifies a category for each requirement in terms of how the requirement should be met. The requirements can be categorized as follows:

1. *Must meet*: The verification tests must be performed, and all applicable requirements must be met by the product.
2. *Use carryover*: The verification compliance received in a previous product program that used the same product design can be accepted. Additional verification tests may not be needed.
3. *Deviation sign-off*: If the testing results show that the product cannot meet the requirement, a deviation must be written (to allow for proceeding without meeting the requirement), approved, and signed-off by the program management.
4. *Does not apply*: The requirement is not relevant to the particular program. Thus, verification tests are not needed.

VALIDATION PLAN AND TESTS

The validation of a complex product also requires a comprehensive validation plan. The validation plan is based on an agreed upon set of tests and evaluations with inputs and product expectations (e.g., quality) of a number of parties (e.g., users, experts, engineers and market researchers from the company management and purchasing organizations, and government agencies responsible for making and enforcing regulatory requirements on the product).

Table 22.1 presents an evaluation test plan for an automotive product. The test plan is structured to evaluate vehicle attributes by using vehicle product-level tests. For validation tests, the product-level tests are conducted using prototype products or early production units. The tests include a combination of evaluation methods involving (1) engineering verification type objective tests that apply to fully assembled product (i.e., a vehicle) using physical test equipment (e.g., dynamometers); (2) experts (e.g., test drivers, technical experts, and test engineers) who evaluate the product using predefined engineering procedures in an objective manner (i.e., not influenced by individual judgments or biases), and (3) customers who are led to evaluate the product using structured evaluation procedures and provide their subjective impressions (e.g.,

TABLE 22.1
Illustration of a Product-Level Evaluation Plan

Vehicle Attribute	Sub-attribute	Sub-attribute Requirements / sources	Evaluation Method(s)
Package	Seating Package (Driver & passengers)	Accommodation percentiles and interior dimensions. SAE J 1516,J1517, J4004.	Interior coordinate measurements. Customer Ratings.
	Entry/Exit	Head/torso, knee, thigh, foot space requirements. Distances from Sating Reference Point (SgRP).	Interior coordinate measurements. Customer Ratings.
	Luggage/Cargo Package	Luggage volume requirements. Floor ht to ground.	Interior coordinate measurements. Customer Ratings.
	Fields of View-Visibility	Wiper/defroster zones, mirror fields, pillar obscuration	Interior coordinate measurements. Customer Ratings.
	Powertrain Package	Engine, transmission and drivetrain envelopes.	Interior coordinate measurements. Eng. Tests.
	Suspensions & Tires Package	Suspension and tire envelopes.	Interior coordinate measurements. Customer Ratings.
	Other Mech Package	Space requirements for fuel tank, electrical, lighting, climate control, etc., systems. FMVSS 108 requirements.	Interior coordinate measurements.
Controls & Displays	Locations-Layout	SAE J1138. Ergonomic requirements.	Interior coordinate measurements.
	Hand & Foot Reach	SAE J287. SAE J1516 and J4004.	Interior coordinate measurements.
	Visibility and Obscurations	FMVSS 111, SAE J1050, J902,903.	Interior coordinate measurements.
	Operability	Ergonomic guidelines, SAE J1139	Ergonomics scorecard.
Safety	Front Impact	FMVSS 204, 208, 212 & 219 requirements.	CAE methods. Sled tests with crash dummies.
	Side Impact	FMVSS 214 requirements.	Engineering tests.

TABLE 22.1 (Continued)
Illustration of a Product-Level Evaluation Plan

Vehicle Attribute	Sub-attribute	Sub-attribute Requirements / sources	Evaluation Method(s)
	Rear Impact	FMVSS 301 and 303 requirements..	CAE tests. Eng. Tests.
	Roof Crush	Deformation requirements.	CAE and Eng. Tests
	Sensors, Belts & Airbags	Anchorage and dummy tests.	CAE and Eng. Tests.
	Other Safety Features	FMVSS 108, SAE lighting stds.	Photometric & Eng. Tests.
Styling/ Appearance	Exterior-- shape, proportions, etc.	Exterior design guidelines.	Exterior surface measurements. Customer ratings.
	Interior-- I/P, Console, trim, etc.	Interior design guidelines	Interior surface measurements. Customer ratings.
	Luggage/Cargo/ Storage	Customer requirements.	Customer ratings.
	Underhood Appearance	Design guidelines.	Customer ratings.
	Color/Texture Mastering	Color and texture masters.	Customer ratings.
	Craftsmanship	Craftsmanship guidelines.	Expert and customer ratings. Measurements of mating edges, surfaces and surface finish.
Thermal & Aerodynamics	Aerodynamics	Aero forces, Coefficient of drag & noise requirements.	CAE and wind tunnel testing.
	Thermal Management	Temperature guidelines	CAE and heat management tests.
	Water Management	Leak test requirements	Water and air leak tests.
Performance & Drivability	Performance Feel	0–60 mph time. Eng. Requirements	Experts and Customer ratings.
	Fuel Economy	EPA/NHTSA requirements	EPA test procedures
	Long Range Capabilities	Eng. Requirements.	Field tests.
	Drivability	Eng. Requirements.	Field tests.

(continued)

TABLE 22.1 (Continued)
Illustration of a Product-Level Evaluation Plan

Vehicle Attribute	Sub-attribute	Sub-attribute Requirements / sources	Evaluation Method(s)
	Manual Shifting	Eng. Requirements.	Experts and Customer ratings.
Vehicle Dynamics	Ride	Eng. Requirements.	Experts and Customer ratings.
	Steering and Handling	Eng. Requirements.	Experts and Customer ratings.
	Braking	FMVSS 105 requirements.	Field tests.
Noise, Vibrations & Harshness (NVH)	Road NVH	Eng. Requirements.	Sound measurements. Field tests.
	Powertrain NVH	Eng. Requirements.	Sound measurements. Field tests.
	Wind Noise	Eng. Requirements.	Sound measurements. Field tests.
	Electrical/ Mechanical	Eng. Requirements.	Field tests.
	Brake NVH	Eng. Requirements.	Field tests.
	Squeaks & Rattles	Eng. Requirements.	Field tests. Customer ratings.
	Pass by Noise	Eng. Requirements.	Sound measurements. Field tests.
Interior Climate Comfort	Heater Performance	Eng. Requirements.	Field tests. Customer ratings.
	A/C Performance	Eng. Requirements.	Field tests. Customer ratings.
	Water Ingestion	Eng. Requirements.	Field tests.
Security	Vehicle Theft	Eng. Requirements.	Expert evaluations.
	Contents/Component Theft	Eng. Requirements.	Expert evaluations.
	Personal Security	Eng. Requirements.	Expert evaluations.
Emissions	Tailpipe emissions	EPA requirements.	Dynamometer and field tests.
	Vapor Emissions	EPA requirements.	Dynamometer and field tests
	On-board diagnostics	EPA requirements.	Dynamometer and field tests
Weight	Body	Design assumptions.	CAE weight predictions & measurements.

TABLE 22.1 (Continued)
Illustration of a Product-Level Evaluation Plan

Vehicle Attribute	Sub-attribute	Sub-attribute Requirements / sources	Evaluation Method(s)
	Chassis	Design assumptions.	CAE weight predictions & measurements.
	Powertrain	Design assumptions.	CAE weight predictions & measurements.
	Climate control	Design assumptions.	CAE weight predictions & measurements.
	Electrical	Design assumptions.	CAE weight predictions & measurements.
Cost	Cost to the customer	Product planning assumptions.	Cost prediction programs.
	Cost to the company	Product planning assumptions.	Cost prediction programs.
Customer Life Cycle	Purchase & Service Experience	Marketing assumptions.	Historic data and customer feedback.
	Operating Experience	Marketing assumptions.	Customer feedback.
	Life Stage Changes	Marketing assumptions.	Customer feedback.
	System Upgrading	Marketing assumptions.	Customer feedback.
	Disposal/ Recyclability	Recycling requirements	Material tracking.
Product/Process Compatibility	Reusability	Reusability requirements.	Field data.
	Commonality	Commonality guidelines.	Analysis of Component database.
	Carryover	Tooling budget.	Analysis of Component database.
	Complexity	Manufacturing budget.	Analysis of Component database.
	Tooling/Plant Life Cycle	Manufacturing strategy and budget.	Analysis of plant tooling database.

Note: CAE, computer-aided engineering; EPA, U.S. Environmental Protection Agency; NHTSA, National Highway Traffic Safety Administration.

ratings on a number of product characteristics, their preferences to the products or product features, comments on what they liked or disliked, and reasons to support their ratings or comments).

The underlying goal in this type of combination of evaluation methods involving physical tests, experts, and customers is to complete the verification and validation within limited time and budget and achieve a high level of objectivity and customer representation. The selection of combinations of objective and subjective evaluations and questions to be included (e.g., who performs the subjective tests: experts versus customers) are usually determined by the group of decision makers involving engineers, managers, and market researchers who are usually guided by the organizations policies, quality manuals or quality management system, and government regulations. In some cases, the decision makers can be experts, managers, users/operators from the organization that purchases the product, a third party (like a registrar for ISO 9000 certification or independent testing agencies), or the government agencies that enforce regulations on the product (e.g., airline companies that purchase airplanes and the Federal Aviation Administration rules and requirements can help in determining validation procedures for an airplane. Similarly, the automotive products are evaluated by users, consumer evaluation organizations, companies purchasing the automotive products and the National Highway Traffic Safety Administration which develops and enforces the federal rules).

It should be noted that the CAE tests are conducted for verification testing. They are not performed for validation testing as actual products are available for these tests. The CAE tests do not use actual physical products they do not represent manufacturing variations and assembly issues (e.g., fit, finish, loose parts, and unpredictable vibrations) related to final built entities. Dimensions and coordinate measurement tests are performed on test samples (used for verification and validation testing) to ensure that test samples (i.e., entities at different levels from component to the whole product) meet at the dimensions and tolerances specified in the CAD drawings and models.

VEHICLE ERGONOMICS EVALUATIONS

The verification tests involved ergonomic evaluations generally include the following:

1) Checking dimensions of driver and passenger positions in the vehicle space (e.g., seating reference point [SgRP] locations).
2) Ensuring that SgRPs of the driver and passengers are located within specified tolerances by using SAE J826 H-point machine (HPM) and/or SAE J4002 H-point device (HPD) in physical mock-ups in package evaluation buck and programmable vehicle bucks (see Volume 1, Chapters 4 and 5).
3) Verifying that controls and displays meet controls and display guidelines (e.g., hand reach, visibility and legibility of displays, pedal and steering wheel locations, SAE J1138 and J1139, FMVSS 101) (see Volume 1, Chapter 7).
4) Verifying that field of view requirements (e.g., FMVSS 103, 104, 111) are met (see Volume 1, Chapter 8).

5) Verifying that entry/exit related vehicle dimensions are within specifications. Selected customers are asked to participate in entry and exit evaluations of the interior package buck to evaluate dimensions and locations of items (e.g., door handles) related to entry/exit (see Volume 1, Chapter 10).

6) All comfort and convenience items (e.g., seat controls, cupholders, handles, storage spaces) meet ergonomics guidelines (see Volume 1, Chapter 7).

7) All exterior service related items (e.g., hood opening control, engine service, fuel filler cap, spare tire and jack storage and operation) should meet ergonomics guidelines (see Volume 1, Chapter 11).

8) Selected features or systems such as climate control, infotainment, navigation, instrument clusters should be evaluated for operability, workload and ease of use by conducting laboratory and/or driving tests (see Chapters 15, 16 and 19).

The validation tests involved ergonomic evaluations primarily include the following:

1) Drive tests with interviewers: The interviewers will lead each customer through a systematic procedure to perform a number of tasks and driving maneuvers and the responses of the customers to questions asked by the interviewers are recorded. The tasks will include operation and use of controls and displays related to interior and exterior tasks and assessing spaces (e.g., headroom, legroom, shoulder room), seating and thermal comfort. The responses obtained through ratings on scales will be analyzed by computing measures such as average ratings, percentage of responses in preferred range (e.g., ratings of 8 or above on 10-point scales), percentage of responses in non-preferred range (e.g., ratings of 3 or below on 10-point scales), percentage of items and item locations disliked, and so forth (see Chapter 19).

2) Naturalist drive tests: The customer will be provided test vehicles installed with selected vehicle systems and components to evaluate under their normal driving conditions over one to a few days and their responses will be recorded on an evaluation form (including issues covered in Volume 1, Chapters 5 through 12).

3) Comparisons with benchmarked vehicles: The subjects will be asked to drive the prototypes of the new vehicle along with selected benchmarked vehicles (e.g., latest models of leading competitors in the same market segment) on selected routes, and their responses to the vehicles will be collected on a number of features. Methods such as rating on a scale, Thurstone's method of paired comparisons and analytical hierarchical method can be used to rank order the vehicles on different measures (see Chapter 19).

CONCLUDING REMARKS

The design teams and their management personnel are responsible for ensuring that their design meets all applicable requirements on the product and the product meets all the customer needs. Thus, evaluations must be conducted to ensure that each entity or the product meets its applicable requirements. The evaluation outcomes should provide support or proof of compliance of the requirements to eliminate future problems. The feedback received from the evaluations is very important to program management. If certain requirements are not met, then other alternatives (e.g., redesign, retest, or procuring the failed entity from another supplier) need to be immediately considered. Such failures put a lot of pressure on the design teams and the management. If the failed requirement is mandatory by a federal standard or a regulation, then the design must be modified and retested until its compliance to the requirements is verified. A failure to meet a requirement can increase financial risks, reduce customer confidence in the product, and degrade the reputation of the manufacturer. An injury caused by a product defect resulting from noncompliance of a requirement can also suggest that the manufacturer was negligent and thus supports the plaintiff's claim in a product liability case. On the contrary, meeting all the applicable requirements will confirm that the right product was developed, and it will boost the morale of the design team and the program management.

REFERENCES

Bhise, V. D. 2012. *Ergonomics in the Automotive Design Process*. Boca Raton, FL: The CRC Press.

Chapanis, A. 1959. *Research Techniques in Human Engineering*. Baltimore, MD: The Johns Hopkins Press.

Kolarik, W. J. 1995. *Creating Quality—Concepts, Systems, Strategies, and Tools*. New York, NY: McGraw-Hill.

Pew, R.W. 1993. *Experimental Design Methodology Assessment*. BBN Report no. 7917. Bolt Beranek & Newman, Inc., Cambridge, MA.

Richards, A. and V. Bhise. 2004. *Evaluation of the PVM Methodology to Evaluate Vehicle Interior Packages*. SAE Paper no. 2004-01-0370. Also published in SAE Report SP-1877, SAE International, Inc., Warrendale, PA.

Satty, T. L. 1980. *The Analytic Hierarchy Process*. New York, NY: McGraw-Hill.

Thurstone, L. L. 1927. The Method of Paired Comparisons for Social Values. *Journal of Abnormal and Social Psychology*, 21: 384–400.

Zikmund, W.G. and B. J. Babin. 2009. *Exploring Market Research*. Ninth Edition. Independence, KY: Boston, MA: Cengage Learning.

23 Costs and Benefit Considerations and Models

INTRODUCTION

One of the key objectives of a product-producing organization is to make money unless the organization is a nonprofit or a government entity. The revenues generated by selling its products and costs of developing, producing, distributing, and maintaining the products are important to all organizations. A primary goal of the organizations is thus to maximize the revenues minus the costs over the entire life cycle of the product – from its conception to product disposal. At the early stages, accurate estimates of the costs are required to develop a budget for the product program and to get it approved. The actual expenditure of costs should be continuously compared with the budgeted costs to ensure that the program is meeting its budgetary requirements. Differences between the budgeted costs and actual costs may signal over or under expenditures or errors in estimating the budgeted costs.

The costs are estimated by breaking down a large product program into a series of manageable tasks. Experienced cost estimators, based on the work content in each task and availability of cost information from previously conducted similar tasks and adjustments for the prevailing and future economic and technological conditions, usually develop time and cost estimates to complete the tasks. The costs of all tasks are then added along with some allowances for errors, interest, inflation, and other unknown or unforeseen problems. The project cost estimates are also refined several times in the program as some of the less predictable tasks and unknown issues (e.g., technology development) are resolved or better understood.

The costs are incurred over time. The costs during early product development are nonrecurring; that is, they do not recur or are one-time type of costs associated with product concept development, product design, detailed engineering, testing, building tools, and facilities. Once production begins, the costs associated with purchasing raw materials, parts purchased from suppliers, plant running costs, direct labor costs, insurance costs, and so forth, are proportional to the volume of products manufactured. The cumulative monetary needs decrease as products are sold and the revenues are generated.

The objective of this chapter is to understand different types of costs associated with the various tasks involved during the product life cycle, learn how to conduct cost-benefit analysis and, most importantly, study how the ergonomics engineers

DOI: 10.1201/9781003485605-10

need to deal with costs and benefits considerations in creating ergonomically superior automotive products.

Creating ergonomically superior products should not involve increasing costs, especially when the ergonomic considerations are provided to the design team early during the concept development phase. For example, the ergonomic wants related to spacious occupant compartment with larger headroom, legroom and shoulder room, easy entry and exit into the vehicle, larger fields of view (e.g., command seating position), controls located within reach zones, displays that are unobstructed and legible to drivers over 65 years old, easy to load and unload from luggage, easy to find and service items from exterior, and so forth would not increase design costs. But the size of the product should affect its weight and power requirements and, hence, increase its cost. Development of new features or changes in features and in-vehicle devices (e.g., navigation system, driver assistance systems) may not increase vehicle costs unless new feature content is added (e.g., additional sensors, actuators and microprocessors, power seats with memory, more comfortable seating materials, adaptive suspensions, programmable instrument clusters, multifunction controls). The costs of such new features need to be justified against additional benefits in terms added comfort, convenience, ease, safety, performance, reduced operation time, and so forth. Converting all the costs and benefits to the present values of monetary units is generally the best way to convince the decision makers (i.e., managers in product development) to incorporate ergonomic changes in a new vehicle. Benefits can be estimated from the increase in vehicle prices the customer is willing to pay for added features that improve functionality, comfort and safety. Accounting for the above factors related to implementation of ergonomic considerations in cost-benefit analysis is discussed in the later part of this chapter.

Cost-benefit analysis is a useful method for decision-making. The analysis requires understanding how the costs and benefits are incurred with each alternative involved in the problem being solved. The benefits generated from the use of a system should be higher than the costs associated with acquiring, operating and maintaining the system. This chapter provides information on how to create a cost-benefit analysis and presents an example of its application to the problem of deciding the best alternative in selecting configuration of a mid-size SUV.

COST-BENEFIT ANALYSIS: WHAT IS IT?

A cost-benefit analysis is a process by which a decision maker can analyze available alternatives and make decisions related to selection of an alternative involving a system or a product. To conduct a cost-benefit analysis, a decision model is first developed by identifying alternatives and outcomes, and by determining benefits and costs of each combination of alternatives and outcomes. The decision is made based on one or more decision principles and maximum values of evaluation measures obtained by a) subtracting the costs from the benefits, and/or b) computing ratios of benefits-to-costs (see decision matrix covered in Volume 1, Chapter 3).

WHY USE COST-BENEFIT ANALYSIS?

Cost-benefit analysis is an objective method of decision-making. However, to determine the alternatives, outcomes, and costs and benefits (associated with each combination of alternatives and outcomes), the cost-benefit analysis relies on the abilities of the analyst and/or the team involved in its formulation of the problem and estimation of values of the variables included in the analysis.

Organizations rely on cost-benefit analysis to support decision-making because it provides an objective and evidence-based view of the issues being evaluated – without the influences of opinions, politics, or biases of decision makers and other individuals within and outside the organization who may directly or indirectly be affected by the decision. By providing quantitative estimates of the consequences of a decision, the cost-benefit analysis is an invaluable tool in developing business strategy, evaluating a proposal, or making resource allocation or purchase decisions. In business, government, finance, and even the nonprofit world, the cost-benefit analysis offers unique and valuable insights. Some examples of its applications are:

a) Comparing project/program proposals
b) Deciding whether to pursue a proposed project
c) Determining combinations of features to be incorporated in a proposed product
d) Evaluating alternate locations for a new product manufacturing plant
e) Weighing different investment opportunities
f) Measuring social benefits of proposed changes in regulations
g) Appraising the desirability of suggested policy alternatives
h) Assessing change initiatives
i) Determining effects of economic or political change on stakeholders and participants

Steps Involved in Cost-Benefit Analysis
While there is no "standard" format for performing a cost-benefit analysis, there are certain core steps in such analyses. The five basic steps to performing a cost-benefit analysis include:

1. Identify alternatives, outcomes, and probabilities of the outcomes
2. Identify costs and benefits so they can be categorized by type (e.g., fixed costs, variable costs, and safety costs)
3. Calculate costs and benefits for each combination of alternative and outcome over the assumed life of the project or initiative at the beginning of the project planning time using the "present value" method described later in this chapter.
4. Select principles to be used in evaluating the alternatives (see Volume 1, Chapter 3)
5. Calculate values of evaluation measures for all alternatives and make recommendations. The most used evaluation measures are a) the net present value (i.e., sum of the present value of benefits – revenues, incentives such

as tax deductions or rebates) over the life of the project minus the sum of the present value of all costs over the life of the project, and b) the benefit-to-cost ratio (i.e., ratio of sum of the present value of benefits over the life of the project to the sum of the present value of all costs over the life of the project. The evaluation measures are computed for each combination of alternatives and outcomes, and the alternative that has the highest value of the evaluation measure is usually selected (or considered further for company management concurrence.

As with any process, it is important to work through all the steps thoroughly and not give in to the temptation to cut corners or base assumptions on opinion or guesses. It is important to ensure that the analysis is as comprehensive as possible (i.e., it covers all costs and benefits incurred over the entire life cycle of the project). Cost-benefit analysis does not require any specific tool to display and calculate costs, benefits, and evaluation measures. However, tabular formats are used to display data and spreadsheets are commonly used to perform calculations.

SOME EXAMPLES OF PROBLEMS FOR APPLICATION OF COST-BENEFIT ANALYSIS

Any decision-making problem in a product or system development and operation area can be analyzed by applying the cost-benefit analysis. Some examples of problems in the automotive systems area that involve computations of costs and benefits associated with different alternatives considered in problem solving are described below:

1. *Select a vehicle concept for developing a new vehicle*: Alternatives considered here are different vehicle concepts created by different vehicle design studios for a specified vehicle type (e.g., sports car, SUV, pickup truck, passenger car). The costs involved are for product development, building facilities, tools and equipment for production and assembly, purchasing components and systems from suppliers, manufacturing and assembly of vehicle components, marketing, sales and so forth. The benefits will be the revenue generated from selling the produced vehicles.
2. *Selecting supplier for a vehicle component or a vehicle system*: An auto manu-facturer does not generally produce all the components and systems that are needed to assemble a vehicle. Many suppliers have the capability to produce and supply the needed components or systems. The auto manufacturer's pur-chasing office is generally responsible for selecting the suppliers based on a number of criteria, such as ability to meet production quantities at required time, price and quality, reputation, expertise and backgrounds of personnel at the supplier's facilities, past experience of working with the supplier, state of manufacturing facilities and equipment in supplier plants, supplier plant loca-tion, and so forth.

TYPES OF COSTS

NON-RECURRING AND RECURRING COSTS

The costs are incurred throughout the life cycle of a product program. The total life cycle costs of a product can be divided into (1) nonrecurring costs and (2) recurring costs.

Nonrecurring Costs

These costs represent expenses and investments that are made during the product development, creation of the production systems, and also to retire and dispose of the systems after the product is terminated. These costs are incurred before the beginning of production and at the end of production, that is, retirement (disposal) stages in the life cycle of a product. The early costs incurred to reach operational status of the program include product design, development, and refinement costs. The costs include personnel (salaries and benefits) of the design team as well as for the development of models, prototypes, market research, verification tests, tools and fixtures design and build, plant and facilities, equipment/tooling installation, and prove-out (research and testing). These nonrecurring costs do not vary as a function of the quantity of products produced. Thus, they are also referred as the "fixed costs".

Recurring Costs

These costs continue to occur and recur throughout the production, sales, and service/maintenance of the products. These costs include personnel costs of the production and distribution (direct and indirect labor), parts and materials purchases, plant and equipment operation and maintenance, utilities, insurance, rents, taxes, licenses, marketing and sales costs, warranty costs, and so forth. The recurring costs vary as a function of the quantity of products produced. Thus, they are also referred to as the "variable costs".

Figure 23.1 shows two charts. The top chart shows various costs (which have negative values, as they represent money spent) as they incur as functions of time during various life cycle stages of a typical product program for a manufactured product. The top chart also shows revenues. The revenues have positive values as they represent income. They are only generated after the products are sold. (Note: Revenue = Units sold × Unit price.) The lower chart in Figure 23.1 shows the systems engineering "V" model. The timeline of the "V" model is synchronized with the timeline of the upper cost chart.

For the cost management purposes, all the costs (negative values) and revenues (positive values) are added, and the present value of cumulative cash flow is frequently reviewed and compared with the budgeted cash flow (i.e., predicted revenue minus budgeted costs). Two cumulative cash flow curves are presented in Figure 23.2. Let us assume that the two cumulative cash flow curves are for two alternative product programs. Alternative (1) incurs more costs and also extends over longer duration in the negative cash flow condition than alternative (2). However, the product in alternative (1) generates revenue at a much higher rate than alternative (2). Understanding of

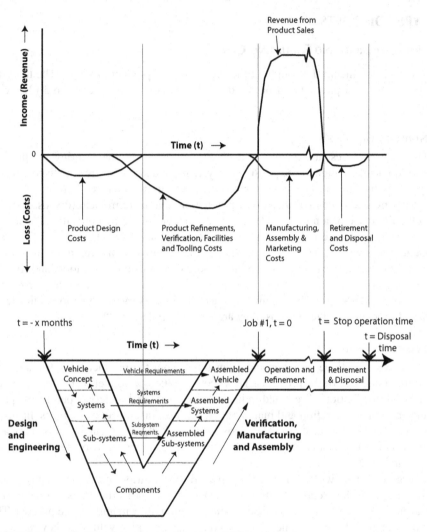

FIGURE 23.1   Costs and revenues in the product life cycle.

the cumulative cash flow curves (i.e., their levels and timings) is especially important before committing to an alternative.

REVENUES

Revenues build up over time as the product is sold. The generated revenues (positive values) are tracked and added to the total costs (negative values). The revenues are also affected by a number of factors such as changes in product volumes due to obsolescence of older products and emergence of new trends in design and technologies, introduction and availability of new products by the competitors, and changes

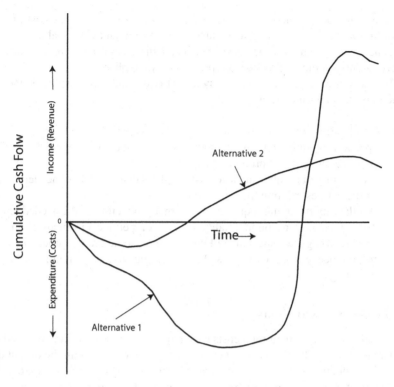

FIGURE 23.2 Cumulative cash flow curves for two alternative product programs.

in economic conditions (e.g., state of economy, interest, inflation, and currency exchange rates).

MAKE VERSUS BUY DECISIONS

Most product-producing organizations do not produce all the entities (i.e., systems, subsystems, or components) of the product within their organization. Many of the entities are purchased from other organizations (i.e., suppliers). Typically, standardized components that are common across many similar products are made by different organizations. Some examples of standardized components are fasteners (such as nuts, bolts, rivets, springs, clips, and pins), electrical and electronic components (e.g., wiring harnesses, switches, resistors, transistors, and microprocessors), plumbing supplies (e.g., valves, pipes, hoses and connectors), and so on. Some special components that require unique manufacturing processes and specialized systems, machines, or equipment are also purchased from suppliers with specialized capabilities. For example, major automotive manufacturers typically purchase about 30%–70% of the components (or systems) in the automotive products from their suppliers. The aircrafts companies also rely on suppliers to produce most of their components. For example, none of the commercial aircraft manufacturing companies produce jet engines that contribute about 40%–50% of the cost of an airplane. Similarly, specialized systems such as electronic and electrical systems with components such as microprocessors, sensors, actuators, printed circuit boards, and so forth in most complex products are produced by suppliers.

The decision whether to make or buy an entity depends on many considerations. Some important considerations include:

1. Availability of in-house manufacturing capability and capacity (e.g., specialized equipment and personnel with unique backgrounds and skills to produce the required product volume)
2. Availability of reliable and low-cost suppliers that can deliver needed volumes of the entities meeting the required quality requirements
3. Availability of capital required for internal production of the needed entities
4. Need to maintain confidentiality of the competitive information on future product designs or specialized knowledge on some unique processes needed to produce certain entities within the organization to retain competitive advantage

FIXED VERSUS VARIABLE COSTS

Organizations organize their total costs into two major categories, namely, fixed costs and variable costs. The fixed costs do not increase or decrease with the output quantity (i.e., production volumes) of products produced. The variable costs are a direct function of the output quantities (i.e., the variable costs increase with an increase in output quantity). The nonrecurring costs are treated as the fixed costs and the recurring costs are the variable costs. The cost of any output is the sum of the fixed and the variable costs. The manufacturers should seek to reduce both the fixed and variable costs. However, decreasing the unit cost of output through increasing product volumes is a much sought-after approach as it spreads the fixed costs over a larger volume. Developing and/or using common components that can be shared across a larger number of products (or models and, hence, increasing their component volumes) can reduce the total cost of the components substantially. Table 23.1 shows the effect of product volume on three products, namely, A, B, and C. The product cost was computed by using the following simple formula:

$$\text{Product cost per unit} = \left(\frac{\text{Fixed costs}}{\text{Product volume}} \right) + \text{Variable cost per unit}$$

Table 23.1 shows that the unit cost of product A will decrease from \$15,000 to \$5,001 as the product volume is increased from 100 units to 1 million units. Similarly, the unit cost of product B can decrease from \$105.00 to \$5.01 for product volumes of 100 units to 1 million units, respectively. This shows the importance and power of increasing the product volume in reducing the cost of products.

TABLE 23.1
Effect of Product Volume on Product Cost

			Product Cost ($)					
					Product Volume (units)			
Product	Fixed Costs	Variable Costs/unit	10	100	1000	10,000	1,00,000	10,00,000
A	$10,00,000.00	$5,000.00	$1,05,000.00	$15,000.00	$6,000.00	$5,100.00	$5,010.00	$5,001.00
B	$10,000.00	$5.00	$1,005.00	$105.00	$15.00	$6.00	$5.10	$5.01
C	$1,000.00	$2.00	$102.00	$12.00	$3.00	$2.10	$2.01	$2.00

QUALITY COSTS

To ensure that the product being designed will meet the customer needs and satisfy the customers, the organization must perform a number of tasks such as a) conduct a number of checks, analyses, and evaluations on product characteristics (attributes), b) implement quality control process, c) honor warranty, and d) repair or replace failed components. The costs incurred for such tasks can be grouped into the following four categories (Campanella, 1990):

1. *Prevention costs*: These costs are associated with the information gathered and analyses conducted to ensure that the right product is being designed and the product will meet its customer needs. Some examples of the activities involved in this cost category are market research, benchmarking, product performance analyses, design reviews, supplier reviews and ratings, supplier quality planning, training, quality administration, and process validations.
2. *Appraisal costs*: These costs are related to various appraisals or evaluations conducted to ensure that incoming components and materials and outgoing products will meet quality requirements. Examples of the activities involved in this cost category are purchasing appraisals, maintenance of laboratories with calibrated state-of-the-art testing equipment and trained staff, measurements and tests, inspections, and plant quality audits.
3. *Internal failure costs*: These costs are incurred at the manufacturer's facilities due to product failures during manufacturing, defects observed during testing, troubleshooting and analyzing the failures, rejected and scrapped units (or components), rework, repairs, and so forth.
4. *External failure costs*: These costs are incurred after the product leaves the manufacturer's facilities and is sold to the customers. The costs are due to handling customer complaints, managing returned products, sending replacements, repairing failed products, product recalls, product litigations and liabilities, penalties, lost sales, and so forth.

MANUFACTURING COSTS

The manufacturing costs can be categorized into the following four broad categories:

1. *Costs of parts (components) and subassemblies purchased from the suppliers*: These costs include expenses incurred in purchasing components, assembled systems and subsystems, and standard fasteners from various suppliers.
2. *Costs of parts manufactured internally within the product manufacturer's plants*: These costs are associated with fixed costs for tooling, equipment, and facilities and variable costs associated with purchasing raw materials, expendable tools, processing and operating machines/equipment, inspection, direct labor, coolants, lubricants, utilities, and so on.
3. *Assembly costs*: These include assembly and inspection related to fixed and variable costs of equipment operation (e.g., fixed costs of fixtures and robots needed for assembly; variable costs to program and run the assembly robots and/or equipment), direct labor costs, and associated employee benefits.

4. *Overhead costs*: These costs include expenses related to indirect labor (e.g., administrative and plant maintenance personnel and costs of their benefits), employee training, utilities, insurance, property taxes, equipment dismantling, and so on.

   It should be noted that all of the preceding four categories have fixed and variable cost components.

SAFETY COSTS

The safety-related costs can be categorized into the following four broad categories:

1. *Accident prevention costs*: These costs represent amounts spent by the organization to avoid or prevent accidents, injuries and adverse health effects from occurring. The accident prevention activities typically include safety analyses (e.g., conducting hazard analyses and failure modes and effects analyses); incorporating engineering changes (e.g., process and equipment improvements to reduce probability of accidents, adding safety devices); conducting safety evaluations/tests and safety reviews; providing safety training to employees; providing/installing and maintaining injury and health protection devices (e.g., hard hats, safety glasses, lockout devices, anti-slip walking surfaces, and providing lifting devices to reduce back injuries).
2. *Costs due to accidents*: These costs include losses that an organization incurs due to accidents. Accidents can involve injuries (e.g., medical costs, temporary disability-related costs until an injured person returns to his/her regular job, and costs due to permanent disability), loss of life, damage to facilities and equipment, and/or work stoppages. It should be noted that accident costs are almost always underestimated due to many unreported or unaccounted costs (e.g., loss of production or temporary work slowdowns due to accidents, retraining of replacement workers). In some cases, the incidental costs of accidents have been estimated to be four times as great as the directly accounted costs.
3. *Insurance costs*: This category includes costs to insure (i.e., insurance premiums and workers compensation costs) against losses due to accidents and injuries, fatalities, and property damage (i.e., repairing or replacing damaged equipment).
4. *Product liability costs*: These are costs incurred in the product liability cases resulting from injuries caused by the product due to defects in the products. These costs include costs to defend cases (e.g., fees charged by lawyers and experts), and compensation or settlement charges paid to the plaintiff, penalties, and fines. (See Chapter 26 for more information on product liability.)

PRODUCT TERMINATION COSTS

These costs are incurred after the decision is made to terminate the production of the product. These costs include the following:

1. Costs of selling discontinued products at discounted prices or with sales incentives

2. Costs of lost sales of new products due to some customers purchasing the discontinued products at the discounted prices
3. Plant and equipment write-down costs
4. Plant shutdown, equipment removal, and disposal costs
5. Environmental cleanup and site restoration costs
6. Materials recycling costs
7. Continual service, production, and distribution of spare parts for products in service until they are disposed of

TOTAL LIFE CYCLE COSTS

These costs include a total of all the aforementioned costs from product conception to end of production and disposal (or recycling) of all products from service and facilities closing.

EFFECT OF TIME ON COSTS

As costs are incurred over time, in determining all the aforementioned costs, the effect of time due to factors such as interest rate, inflation rate, and fluctuations in currency exchange rates (if applicable) must be taken into account. Similarly, since the revenues are generated over the selling periods of the products and payments are received over time, the effects of changes in interest rates, inflation, and currency exchange rates should also be considered.

Most complex product programs extend over many years. Therefore, cost computations need to consider the effects of interest and inflation. The computations can be made by using the following variables and the formula.

Let P = value at a time assumed to be the present (called the present value)
i = combined annual interest and inflation rate = $i_r + i_f$
i_r = annual interest rate
i_f = annual inflation rate
n = number of annual interest periods
F = future value after n periods

With the annual compounding of the combined interest and inflation, the relationship between P and F is as follows (Blanchard and Fabrycky, 2011):

$$F = P(1+i)^n \text{ or } P = F\left\{\frac{1}{(1+i)^n}\right\}$$

Using the preceding formula, the value of $100 today will be $128 in 5 years at 5% combined annual interest and inflation rate. (Note: $128 = 100 (1 + 0.05)^5$) This means that $128 spent 5 years from now will be equivalent to $100 today assuming 5% rate of combined interest and inflation.

For a program extending over many periods, the present value of revenues minus the present value of costs can be computed for each period; and the net present values for each period can be summed over the entire duration of the program to obtain the present value of the cumulative cash flow. The present value is generally computed at the beginning of the product program to provide the management with an estimate of cash flow over the life of the program.

BENEFITS ESTIMATION

Benefits include income that an organization earns by selling its products and services, investment income and other incentives it receives through tax breaks, rebates and special arrangements through federal, state or local governments.

The revenues are the income from selling products and services, which are computed by multiplying units sold by the selling price of the unit. Investment income is the interest and dividends received and appreciated value of investments made by the product producing organization. Incentives typically include reduction in local, state or federal taxes, special lower interest rates received through government agencies for locating facilities (e.g., a production or assembly plant) of the organization. Other examples of incentives are the value of power buyback provisions paid by the states to companies using renewable energy, energy efficient equipment or reduction of pollutants.

PROGRAM FINANCIAL PLAN

AN EXAMPLE: AUTOMOTIVE PRODUCT PROGRAM CASH FLOW

This section presents a simplified cash flow model of an automotive product program. The cash flow analysis covers a 100-month period – from 40 months before Job #1 to 60 months after Job #1. In the automotive industry, Job #1 represents the time at which the first production vehicle rolls out of the assembly plant.

The assumptions used for the costs and revenue computations were as follows:

Program milestones: Program kicks off at –40 months (–40 months represents 40 months before Job #1)

1. Product development team formation begins at –39.5 months
2. Strategic intent confirmation at –34 months
3. Hard-points freeze at –29 months
4. Feasibility sign-off at –27.5 months
5. Program approval at –26 months
6. Surface freeze at –24 months
7. Appearance approval at –19 months
8. Early prototype vehicles available for testing at –14 months
9. Early production prototype vehicles available for testing at –9 months
10. Final prototype vehicles available at –5 months
11. Final sign-off at –2 months

12. Job #1 at 0 month
13. Postproduction at 60 months after Job #1

Cost inputs:

1. Salaried head count cost (i.e., salary plus benefits) = $8,000/month
2. Manufacturing personnel cost = $30/h
3. Manufacturing plant operation = 2 shifts/day, 25 days/month
4. Purchased parts and plant overhead costs = $12,000/vehicle
5. Marketing cost = $2,000/vehicle

These values were used to compute evaluation measures for decision matrix presented in Volume 1, Chapter 3 (Table 3.3). In this problem it was assumed that the automotive manufacturer wants to select a design for its future mid-size SUV. The alternatives and outcomes considered in the decision matrix are as follows.
The manufacturer is considering the following five alternatives:

A_1 = Develop a new mid-size SUV with a 250 HP 2.0 L ICE and 2-row seating
A_2 = Develop a new mid-size SUV with a hybrid powertrain with 300 HP 2.0 turbo-boost plus an electric motor and 2-row seating
A_3 = Develop a new mid-size SUV with a 320 HP 2.3 L turbo-boost plus an electric motor (hybrid) and 3-row seating
A_4 = Develop a new mid-size SUV with a 335 HP electric powertrain with twin motors (full-electric) with 3-row seating
A_5 = Improve an existing 280 HP 2.3L turbo-boost (ICE) SUV with 2-row seating

Six possible outcomes assumed by the manufacturer are as follows:

O_1 = Economy does not change -- oil prices remain low, and the battery technology does not improve
O_2 = Economy improves by 5%, oil prices remain low, and battery technology does not improve
O_3 = Economy degrades by 5%, oil prices increase by 30% and the battery technology does not improve
O_4 = Economy does not change; oil prices remain low and battery technology improves by 50%
O_5 = Economy improves by 5%, oil prices remain low and battery technology improves by 50%
O_6 = Economy degrades by 5%, oil prices increase by 30% and the battery technology improves by 50%

Table 23.3 presents a spread sheet showing computations of various costs and revenues during the first 100 months of the automotive product program. (Note: This vehicle program was defined as alternative 1 [A1] and outcome 1 [O1] presented in Volume

TABLE 23.2

Values of Vehicle Production, MSRP and Purchased Parts Used for Cash Flow Analysis

Alternative	Maximum Monthly Vehicle Production							Purchased
Outcomes ->	O1	O2	O3	O4	O5	O6	MSRP ($)	Parts ($)
A1	4000	4200	3800	4400	5500	5400	40000	12000
A2	4500	4725	3325	4950	6190	6075	43000	15000
A3	3500	3675	3325	3850	4815	4725	46000	18000
A4	3000	3150	2850	3300	4125	4050	51000	22000
A5	5000	5250	4750	5500	6875	6750	38000	13000

1, Table 3.3. The values of the evaluation measures shown in the decision matrix presented in Table 3.3 were computed by using the values provided in Table 23.2). Table 23.2 presents assumed values of monthly vehicle sales, manufacturer's suggested vehicle price (MSRP) and cost of purchased price of parts and systems per vehicle produced used in exercising the model (spreadsheet in Table 23.3).

Figure 23.3 presents the present value cumulative costs, revenues and cash flow curves for the vehicle program. The cumulative cash curve was obtained by summing all the costs (negative values) and revenues from product sales (positive values). It should be noted that the present values of the costs and the revenue values in Figure 23.3 and Table 23.3 were computed by using a 3% discount (interest) rate and no inflation.

Table 23.3 shows the point of maximum cumulative expenditure occurred at Job #1 (at month 0). The maximum cumulative expenditure in the program was $1.64 billion at Job #1. The maximum cash flow at 24 and 60 months after Job #1 was $155.2 million and $2.4 billion dollars, respectively.

EFFECT OF ERGONOMIC DESIGN CHANGES ON VEHICLE PROGRAM FINANCES

The vehicle program financial plan (spreadsheet) shown in Table 23.3 can be used to determine effects of costs due to ergonomic design changes or improvements on the vehicle program. The spreadsheet can be reset by changing the cost of the parts, materials and overhead (column K) to represent the costs due to ergonomic changes. If these changes affect the price of the vehicle, it can be changed in the bottom left side (last line) of the spreadsheet.

Table 23.4 presents results of a sensitivity analysis exercise conducted using the spreadsheet model by combination of adding the costs of ergonomic design changes of 0, 10, 50, 100 and 1000 dollars per vehicle and increasing the vehicle price by 0, 10, 50, 100 and 1000 dollars. The outputs of the sensitivity analysis are shown in Table 23.4 by the following two present value (PV) based variables: a) ratio of

TABLE 23.3
Costs, Revenues and Cash Flows in an Automotive Product Program

A Months from Job#1	B Product Development Headcount	C Prod. Dev. Manpower Costs	D Services and Supplies Costs	E Facilities and Tooling Costs	F Product Development Costs Subtotal	G Present Value of Development Costs Subtotal	H Mfg Headcount	I Mfg. Manpower Costs	J Number of Vehicles Produced	K Parts, Materials & overhead Costs	L Manufacturing Costs Subtotal	M Sales and Marketing Costs	N Total Costs	O Revenue from Vehicle Sales	P Present Value of Total Cost	Q Present Value of Revenue from Vehicle Sales	R Present Value of Cumulative Total Cost	S Present Value of Cumulative Total Revenue	T Present Value Cash Flow
-40	50	400,000	140,000		540,000	540,000							540,000		540,000		540,000	0	-540,000
-39	100	800,000	280,000		1,080,000	1,077,307							1,080,000		1,077,307		1,617,307	0	-1,617,307
-38	200	1,600,000	560,000		2,160,000	2,149,240							2,160,000		2,149,240		3,766,547	0	-3,766,547
-37	500	4,000,000	1,400,000		5,400,000	5,359,702							5,400,000		5,359,702		9,126,249	0	-9,126,249
-36	800	6,400,000	2,240,000		8,640,000	8,554,137							8,640,000		8,554,137	0	17,680,386	0	-17,680,386
-35	1,000	8,000,000	2,800,000		10,800,000	10,666,007							10,800,000		10,666,007	0	28,346,393	0	-28,346,393
-34	1,000	8,000,000	2,800,000		10,800,000	10,639,408							10,800,000		10,639,408	0	38,985,801	0	-38,985,801
-33	1,200	9,600,000	3,360,000		12,960,000	12,735,451							12,960,000		12,735,451	0	51,721,252	0	-51,721,252
-32	1,200	9,600,000	3,360,000		12,960,000	12,703,692							12,960,000		12,703,692	0	64,424,944	0	-64,424,944
-31	1,200	9,600,000	3,360,000		12,960,000	12,672,012							12,960,000		12,672,012	0	77,096,956	0	-77,096,956
-30	1,200	9,600,000	3,360,000		12,960,000	12,640,411							12,960,000		12,640,411	0	89,737,366	0	-89,737,366
-29	1,200	9,600,000	3,360,000		12,960,000	12,608,889							12,960,000		12,608,889	0	102,346,255	0	-102,346,255
-28	1,200	9,600,000	3,360,000		12,960,000	12,577,445							12,960,000		12,577,445	0	114,923,700	0	-114,923,700
-27	1,200	9,600,000	3,360,000		12,960,000	12,546,080							12,960,000		12,546,080	0	127,469,780	0	-127,469,780
-26	1,200	9,600,000	3,360,000		12,960,000	12,514,793							12,960,000		12,514,793	0	139,984,573	0	-139,984,573
-25	1,200	9,600,000	3,360,000	40,000,000	52,960,000	51,013,164							52,960,000		51,013,164	0	190,997,736	0	-190,997,736
-24	1,200	9,600,000	3,360,000	40,000,000	52,960,000	50,885,949							52,960,000		50,885,949	0	241,883,685	0	-241,883,685
-23	1,200	9,600,000	3,360,000	40,000,000	52,960,000	50,759,051							52,960,000		50,759,051	0	292,642,736	0	-292,642,736
-22	1,200	9,600,000	3,360,000	40,000,000	52,960,000	50,632,470							52,960,000		50,632,470	0	343,275,206	0	-343,275,206
-21	1,200	9,600,000	3,360,000	40,000,000	52,960,000	50,506,204							52,960,000		50,506,204	0	393,781,410	0	-393,781,410
-20	1,200	9,600,000	3,360,000	40,000,000	52,960,000	50,380,254							52,960,000		50,380,254	0	444,161,664	0	-444,161,664
-19	1,200	9,600,000	3,360,000	40,000,000	52,960,000	50,254,617							52,960,000		50,254,617	0	494,416,281	0	-494,416,281
-18	1,200	9,600,000	3,360,000	40,000,000	52,960,000	50,129,294							52,960,000		50,129,294	0	544,545,575	0	-544,545,575
-17	1,200	9,600,000	3,360,000	40,000,000	52,960,000	50,004,283							52,960,000		50,004,283	0	594,549,858	0	-594,549,858
-16	1,200	9,600,000	3,360,000	40,000,000	52,960,000	49,879,584							52,960,000		49,879,584	0	644,429,443	0	-644,429,443
-15	1,200	9,600,000	3,360,000	40,000,000	52,960,000	49,755,196							52,960,000		49,755,196	0	694,184,639	0	-694,184,639
-14	1,200	9,600,000	3,360,000	40,000,000	52,960,000	49,631,119							52,960,000		49,631,119	0	743,815,757	0	-743,815,757
-13	1,200	9,600,000	3,360,000	40,000,000	52,960,000	49,507,350							52,960,000		49,507,350	0	793,323,108	0	-793,323,108
-12	1,200	9,600,000	3,360,000	40,000,000	52,960,000	49,383,890							52,960,000		49,383,890	0	842,706,998	0	-842,706,998
-11	1,200	9,600,000	3,360,000	40,000,000	52,960,000	49,260,739							52,960,000		49,260,739	0	891,967,737	0	-891,967,737
-10	1,200	9,600,000	3,360,000	40,000,000	52,960,000	49,137,894							52,960,000		49,137,894	0	941,105,630	0	-941,105,630
-9	1,200	9,600,000	3,360,000	40,000,000	52,960,000	49,015,355							52,960,000		49,015,355	0	990,120,986	0	-990,120,986
-8	1,200	9,600,000	3,360,000	40,000,000	52,960,000	48,893,123							52,960,000		48,893,123	0	1,039,014,109	0	-1,039,014,109
-7	1,200	9,600,000	3,360,000	40,000,000	52,960,000	48,771,195							52,960,000		48,771,195	0	1,087,785,303	0	-1,087,785,303
-6	1,000	8,000,000	2,800,000	40,000,000	50,800,000	46,665,374							50,800,000		46,665,374	0	1,134,450,677	0	-1,134,450,677
-5	1,000	8,000,000	2,800,000	40,000,000	50,800,000	46,549,001							50,800,000		46,549,001	0	1,180,999,678	0	-1,180,999,678
-4	800	6,400,000	2,240,000	40,000,000	48,640,000	44,458,606							48,640,000		44,458,606	0	1,225,458,284	0	-1,225,458,284
-3	600	4,800,000	1,680,000	40,000,000	46,480,000	42,378,347	500	6,000,000	200	2,400,000	8,400,000	400,000	55,280,000	8,000,000	50,401,786	7,294,036	1,275,860,070	7,294,036	-1,268,566,034
-2	400	3,200,000	1,120,000	40,000,000	44,320,000	40,308,187	600	7,200,000	300	3,600,000	55,120,000	600,000	100,040,000	12,000,000	90,984,454	10,913,769	1,366,844,524	18,207,805	-1,348,636,720
-1	300	2,400,000	840,000		3,240,000	2,939,369	800	9,600,000	500	6,000,000	18,840,000	1,000,000	23,080,000	20,000,000	20,938,470	18,144,254	1,387,782,994	36,352,059	-1,351,430,935
0	200	1,600,000	560,000		2,160,000	1,954,693	1,000	12,000,000	1,000	12,000,000	26,160,000	2,000,000	30,320,000	40,000,000	27,438,094	36,198,014	1,415,221,088	72,550,073	-1,342,671,015
1					0	0	1,000	12,000,000	1,000	12,000,000	24,000,000	2,000,000	26,000,000	40,000,000	23,470,034	36,107,744	1,438,691,122	108,657,817	-1,330,033,305
2					0	0	1,000	12,000,000	1,000	12,000,000	24,000,000	2,000,000	26,000,000	40,000,000	23,411,505	36,017,700	1,462,102,627	144,675,517	-1,317,427,110
3					0	0	1,000	12,000,000	1,600	19,200,000	31,200,000	3,200,000	34,400,000	64,000,000	30,897,977	37,514,020	1,493,000,604	182,189,537	-1,310,811,067
4					0	0	1,000	12,000,000	1,600	19,200,000	31,200,000	3,200,000	34,400,000	64,000,000	30,820,925	57,341,256	1,523,821,529	239,530,792	-1,284,290,737
5					0	0	1,000	12,000,000	1,600	19,200,000	31,200,000	3,200,000	34,400,000	64,000,000	30,744,066	57,198,260	1,554,565,593	296,729,052	-1,257,836,542
6					0	0	1,000	12,000,000	2,800	33,600,000	45,600,000	5,600,000	51,200,000	112,000,000	45,644,497	99,847,336	1,600,210,090	396,576,388	-1,203,633,702
7					0	0	1,000	12,000,000	2,800	33,600,000	45,600,000	5,600,000	51,200,000	112,000,000	45,530,670	99,598,340	1,645,740,760	496,174,729	-1,149,566,031
8					0	0	1,000	12,000,000	2,800	33,600,000	45,600,000	5,600,000	51,200,000	112,000,000	45,417,127	99,349,966	1,691,157,887	595,524,694	-1,095,633,193
9					0	0	1,000	12,000,000	3,600	43,200,000	55,200,000	7,200,000	62,400,000	144,000,000	55,214,088	127,417,127	1,746,371,975	722,941,821	-1,023,430,154
10					0	0	1,000	12,000,000	3,600	43,200,000	55,200,000	7,200,000	62,400,000	144,000,000	55,076,397	127,099,379	1,801,448,373	850,041,200	-951,407,173
11					0	0	1,000	12,000,000	3,600	43,200,000	55,200,000	7,200,000	62,400,000	144,000,000	54,939,050	126,782,423	1,856,387,423	976,823,623	-879,563,800
12					0	0	1,000	12,000,000	4,000	48,000,000	60,000,000	8,000,000	68,000,000	160,000,000	59,720,177	140,518,063	1,916,107,600	1,117,341,686	-798,765,914
13					0	0	1,000	12,000,000	4,000	48,000,000	60,000,000	8,000,000	68,000,000	160,000,000	59,571,249	140,167,644	1,975,678,848	1,257,509,330	-718,169,518
14					0	0	1,000	12,000,000	4,000	48,000,000	60,000,000	8,000,000	68,000,000	160,000,000	59,422,692	139,818,099	2,035,101,540	1,397,327,429	-637,774,111
15					0	0	1,000	12,000,000	4,000	48,000,000	60,000,000	8,000,000	68,000,000	160,000,000	59,274,506	139,469,425	2,094,376,046	1,536,796,854	-557,579,192
16					0	0	1,000	12,000,000	4,000	48,000,000	60,000,000	8,000,000	68,000,000	160,000,000	59,126,689	139,121,621	2,153,502,735	1,675,918,476	-477,584,260
17					0	0	1,000	12,000,000	4,000	48,000,000	60,000,000	8,000,000	68,000,000	160,000,000	58,979,241	138,774,685	2,212,481,976	1,814,693,160	-397,788,816
18					0	0	1,000	12,000,000	4,000	48,000,000	60,000,000	8,000,000	68,000,000	160,000,000	58,832,161	138,428,613	2,271,314,137	1,953,121,773	-318,192,363
19					0	0	1,000	12,000,000	4,000	48,000,000	60,000,000	8,000,000	68,000,000	160,000,000	58,685,447	138,083,405	2,329,999,584	2,091,205,178	-238,794,406
20					0	0	1,000	12,000,000	4,000	48,000,000	60,000,000	8,000,000	68,000,000	160,000,000	58,539,099	137,739,057	2,388,538,683	2,228,944,235	-159,594,448
21					0	0	1,000	12,000,000	4,000	48,000,000	60,000,000	8,000,000	68,000,000	160,000,000	58,393,116	137,395,568	2,446,931,799	2,366,339,803	-80,591,996
22					0	0	1,000	12,000,000	4,000	48,000,000	60,000,000	8,000,000	68,000,000	160,000,000	58,247,498	137,052,936	2,505,179,297	2,503,392,738	-1,786,558
23					0	0	1,000	12,000,000	4,000	48,000,000	60,000,000	8,000,000	68,000,000	160,000,000	58,102,242	136,711,158	2,563,281,539	2,640,103,896	76,822,357
24					0	0	1,000	12,000,000	4,000	48,000,000	60,000,000	8,000,000	68,000,000	160,000,000	57,957,349	136,370,232	2,621,238,888	2,776,474,128	155,235,241
25							1,000	12,000,000	4,000	48,000,000	60,000,000	8,000,000	68,000,000	160,000,000	57,812,817	136,030,157	2,679,051,704	2,912,504,285	233,452,581
26							1,000	12,000,000	4,000	48,000,000	60,000,000	8,000,000	68,000,000	160,000,000	57,668,645	135,690,929	2,736,720,349	3,048,195,215	311,474,866
27							1,000	12,000,000	4,000	48,000,000	60,000,000	8,000,000	68,000,000	160,000,000	57,524,833	135,352,548	2,794,245,182	3,183,547,763	389,302,581
28							1,000	12,000,000	4,000	48,000,000	60,000,000	8,000,000	68,000,000	160,000,000	57,381,379	135,015,011	2,851,626,562	3,318,562,773	466,936,212
29							1,000	12,000,000	4,000	48,000,000	60,000,000	8,000,000	68,000,000	160,000,000	57,238,264	134,678,315	2,908,864,846	3,453,241,088	544,376,243
30							1,000	12,000,000	4,000	48,000,000	60,000,000	8,000,000	68,000,000	160,000,000	57,095,545	134,342,459	2,965,960,390	3,587,583,547	621,623,156
31							1,000	12,000,000	4,000	48,000,000	60,000,000	8,000,000	68,000,000	160,000,000	56,953,162	134,007,440	3,022,913,553	3,721,590,987	698,677,434
32							1,000	12,000,000	4,000	48,000,000	60,000,000	8,000,000	68,000,000	160,000,000	56,811,134	133,673,257	3,079,724,687	3,855,264,244	775,539,557
33							1,000	12,000,000	4,000	48,000,000	60,000,000	8,000,000	68,000,000	160,000,000	56,669,461	133,339,907	3,136,394,147	3,988,604,151	852,210,004
34							1,000	12,000,000	4,000	48,000,000	60,000,000	8,000,000	68,000,000	160,000,000	56,528,140	133,007,389	3,192,922,287	4,121,611,540	928,689,252
35							1,000	12,000,000	4,000	48,000,000	60,000,000	8,000,000	68,000,000	160,000,000	56,387,172	132,675,699	3,249,309,460	4,254,287,239	1,004,977,779
36							1,000	12,000,000	4,000	48,000,000	60,000,000	8,000,000	68,000,000	160,000,000	56,246,556	132,344,837	3,305,556,016	4,386,632,076	1,081,076,061
37							1,000	12,000,000	3,600	43,200,000	55,200,000	7,200,000	62,400,000	144,000,000	51,485,772	118,813,320	3,357,041,788	4,505,445,397	1,148,403,609
38							1,000	12,000,000	3,600	43,200,000	55,200,000	7,200,000	62,400,000	144,000,000	51,357,379	118,517,028	3,408,399,166	4,623,962,424	1,215,563,258
39							1,000	12,000,000	3,600	43,200,000	55,200,000	7,200,000	62,400,000	144,000,000	51,229,306	118,221,474	3,459,628,472	4,742,183,898	1,282,555,427
40							1,000	12,000,000	3,600	43,200,000	55,200,000	7,200,000	62,400,000	144,000,000	51,101,552	117,926,657	3,510,730,023	4,860,110,556	1,349,380,532
41							1,000	12,000,000	3,600	43,200,000	55,200,000	7,200,000	62,400,000	144,000,000	50,974,116	117,632,576	3,561,704,140	4,977,743,132	1,416,038,992
42							1,000	12,000,000	3,600	43,200,000	55,200,000	7,200,000	62,400,000	144,000,000	50,846,999	117,339,228	3,612,551,138	5,095,082,360	1,482,531,221
43							1,000	12,000,000	3,200	38,400,000	50,400,000	6,400,000	56,800,000	128,000,000	46,168,386	104,041,432	3,658,719,524	5,199,123,792	1,540,404,268
44							1,000	12,000,000	3,200	38,400,000	50,400,000	6,400,000	56,800,000	128,000,000	46,053,252	103,781,977	3,704,772,776	5,302,905,769	1,598,132,993
45							1,000	12,000,000	3,200	38,400,000	50,400,000	6,400,000	56,800,000	128,000,000	45,938,406	103,523,169	3,750,711,183	5,406,428,939	1,655,717,756
46							1,000	12,000,000	3,200	38,400,000	50,400,000	6,400,000	56,800,000	128,000,000	45,823,847	103,266,007	3,796,535,030	5,509,693,946	1,713,158,916
47							1,000	12,000,000	3,200	38,400,000	50,400,000	6,400,000	56,800,000	128,000,000	45,709,573	103,007,488	3,842,244,602	5,612,701,434	1,770,456,831
48							1,000	12,000,000	3,200	38,400,000	50,400,000	6,400,000	56,800,000	128,000,000	45,595,584	102,750,612	3,887,840,186	5,715,452,046	1,827,611,859
49							1,000	12,000,000	2,800	33,600,000	45,600,000	5,600,000	51,200,000	112,000,000	40,997,750	89,682,579	3,928,837,937	5,805,134,624	1,876,296,686
50							1,000	12,000,000	2,800	33,600,000	45,600,000	5,600,000	51,200,000	112,000,000	40,895,512	89,458,931	3,969,733,448	5,894,593,556	1,924,860,108
51							1,000	12,000,000	2,800	33,600,000	45,600,000	5,600,000	51,200,000	112,000,000	40,793,528	89,235,842	4,010,526,976	5,983,829,398	1,973,302,422
52							1,000	12,000,000	2,800	33,600,000	45,600,000	5,600,000	51,200,000	112,000,000	40,691,798	89,013,309	4,051,218,774	6,072,842,706	2,021,623,932
53							1,000	12,000,000	2,800	33,600,000	45,600,000	5,600,000	51,200,000	112,000,000	40,590,322	88,791,330	4,091,809,097	6,161,634,036	2,069,824,940
54							1,000	12,000,000	2,800	33,600,000	45,600,000	5,600,000	51,200,000	112,000,000	40,489,100	88,569,905	4,132,298,196	6,250,203,942	2,117,905,746
55							1,000	12,000,000	2,800	33,600,000	45,600,000	5,600,000	51,200,000	112,000,000	40,388,129	88,349,033	4,172,686,325	6,338,552,975	2,165,866,649
56							1,000	12,000,000	2,800	33,600,000	45,600,000	5,600,000	51,200,000	112,000,000	40,287,411	88,128,711	4,212,973,736	6,426,681,686	2,213,707,950
57							1,000	12,000,000	2,800	33,600,000	45,600,000	5,600,000	51,200,000	112,000,000	40,186,943	87,908,939	4,253,160,680	6,514,590,625	2,261,429,945
58							1,000	12,000,000	2,800	33,600,000	45,600,000	5,600,000	51,200,000	112,000,000	40,086,727	87,689,714	4,293,247,406	6,602,280,339	2,309,032,933
59							1,000	12,000,000	2,800	33,600,000	45,600,000	5,600,000	51,200,000	112,000,000	39,986,760	87,471,037	4,333,234,166	6,689,751,376	2,356,517,210
60							1,000	12,000,000	2,800	33,600,000	45,600,000	5,600,000	51,200,000	112,000,000	39,887,042	87,252,905	4,373,121,208	6,777,004,281	2,403,883,072

Alternative = A1 Outcome = O1
Discount rate for present value calculations = 3
MSRP ($) per vehicle = 40,000
Monthly production = 4,000
Purchased parts ($)/veh= 12,000

Benefit-to-cost ratio = 1
Approximate Annual ROI (%) = 7
Total Program Time (years) = 8

Present value of Revenue/Costs 2

present value of total revenue to present value of total costs, and b) present value of cash flow (total revenue minus present value of total costs). The results show that the highest present value of revenues-to-costs ratio of 1.304 can be obtained by adding zero dollars cost to the incorporate ergonomic changes and increasing the vehicle price by $1000. Whereas adding $1000 cost to incorporate the ergonomic changes

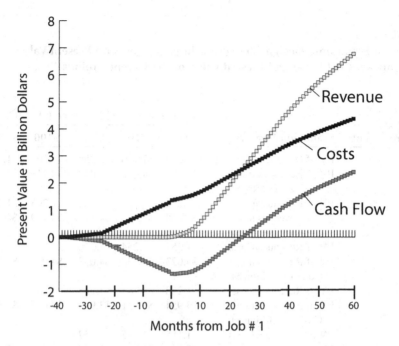

FIGURE 23.3   Costs, revenues and cash flow curves for the vehicle program.

without increasing the vehicle price will reduce the revenues-to-costs ratio to 1.155. If the ergonomic changes are expected to affect the sales volume of the vehicle, estimates of the number of vehicles produced (column J) can be changed with the help of marketing and product planning experts.

The data provided in Table 23.4 show that if the vehicle price is not increased then the present value of the revenues minus the costs of the vehicle program will range from $4.09 to $2.664 billion as the costs of incorporating ergonomic changes is increased from $0.0 to $1000.0 per vehicle. If the vehicle price is increased by $1000, then the revenues minus the costs of the vehicle program will range from $4.802 to $3.373 billion as the costs of incorporating ergonomic changes are increased from $0.0 to $1000.0 per vehicle.

Large complex features involving addition of hardware and software require substantial amount of design, manufacturing and assembly expenditures which can be recovered by increasing the price of the vehicle. The increased price can reduce vehicle sales and the company's market share. Thus, ergonomics engineers must be very careful in understanding the effect of proposing new changes or improvements in the vehicle. The effect of ergonomic changes on customer satisfaction and resulting sales volumes and increase in vehicle costs need to be estimated and evaluated by using a cost-benefit model (e.g., see Table 23.3).

TABLE 23.4
Effect of Ergonomic Design Changes or Improvements on Present Value Revenue-to-Cost Ratio and Present Value of the Revenue minus Cost

Cost of Ergonomic Changes or Improvements	Evaluation Measure	Increase in Vehicle Price (MSRP) by:				
		$0	$10	$50	$100	$1,000
$0	PV of Revenues/Costs	1.256	1.256	1.258	1.26	1.304
	PV of Revenues minus Costs (in Billion $)	$4.094	$4.080	$4.080	$4.118	$4.802
$10	PV of Revenues/Costs	1.258	1.259	1.26	1.263	1.303
	PV of Revenues minus Costs (in Billion $)	$4.078	$4.085	$4.113	$4.149	$4.788
$50	PV of Revenues/Costs	1.254	1.254	1.256	1.258	1.299
	PV of Revenues minus Costs (in Billion $)	$4.023	$4.030	$4.058	$4.094	$4.732
$100	PV of Revenues/Costs	1.248	1.249	1.251	1.253	1.293
	PV of Revenues minus Costs (in Billion $)	$3.954	$3.961	$3.989	$4.025	$4.663
$1,000	PV of Revenues/Costs	1.155	1.155	1.157	1.159	1.196
	PV of Revenues minus Costs (in Billion $)	$2.664	$2.671	$2.699	$2.735	$3.373

PRODUCT PRICING APPROACHES

TRADITIONAL COSTS-PLUS APPROACH

The traditional approach in determining the product price is to add all the costs per unit (of product) and the required profit per unit to come up with the price for the unit. This approach does not provide strong incentives to reduce costs as the profits for the manufacturer are assured. The approach also assumes that customers are willing to pay the price (i.e., it is the producer's market – the producer sets the price without regard to the customers). This approach has worked well in the past when the customers had a limited number of choices in the market for selecting a product.

MARKET PRICE-MINUS PROFIT APPROACH

In this approach, the producer determines the lowest price based on the prices of other similar products sold in the markets, then subtracts his dealer margin and expected profit, and the balance is considered as the target cost for producing the product. The target cost is then divided and assigned to each entity in the product. And all internal and external suppliers are asked to meet their respective target costs by improving the product design, manufacturing processes, and operations.

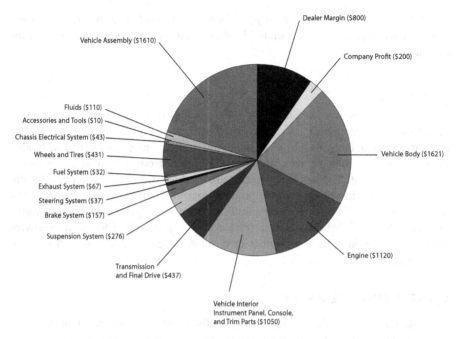

FIGURE 23.4 Low-cost vehicle target cost breakdown.

Source: **Hussain and Randive (2011).**

For example, in determining the price of a low-cost vehicle for the US market, Hussain and Randive (2011) surveyed the prices of low-cost vehicles sold in the US market. They found that the lowest price of small economy vehicles sold in the US market during 2010–2011 was about $10,000. Thus, they set the target manufacturer's retail price of $8,000 (20% below the lowest priced sold in the US market). Assuming the dealer margin of 10% ($800) and the manufacturer's profit of $200 (2.78% of factory cost), they set the target cost at $7,000 per vehicle and then proceeded to develop target cost for each vehicle system (see Figure 23.4). This assumes that they challenged their suppliers to deliver the systems at the target costs. This approach was also used during the development of the Tata Nano, the lowest cost vehicle sold for about INR 100,000 (USD 2,000) in India (Hussain and Randive, 2011).

TRADE-OFFS AND RISKS

Programs and projects involving development of complex products encounter a number of developmental problems and challenges. Many problems involve trade-offs between different attribute requirements and trade-offs between a number of design and manufacturing issues. The costs and timings are directly affected by how the trade-off issues are resolved (see Volume 1, Chapter 3). The design teams deal with these issues constantly during various design stages. Many of these problems are not sufficiently known in the early stages and, hence, the budgets prepared during the early stages need to be constantly reviewed and some changes in target costs and timings may need to be incorporated in subsequent budgets and milestones in the program.

SOFTWARE APPLICATIONS

Many different software applications are available to perform product life cycle costing and to create various reports (e.g., by systems, program phases, months; comparisons with budgeted costs). Many of the applications are integrated with other functions such as management information systems, product planning, and supply chain management. The software systems are used for production scheduling, component ordering, inventory control, product control, shop floor management, cost accounting, and so forth. Some examples of such software systems are manufacturing resource planning (MRP) and enterprise resource planning (ERP). The software systems are available from a number of developers (e.g., SAP, Oracle, Microsoft, EPICOR, and Sage).

CONCLUDING REMARKS

Bringing the right product to the market at the right time and the right price are both very important. The right product requires a design that satisfies the customers. And costs and timings are important parameters used by the program and project managers to evaluate and control their progress. Both these parameters affect the profitability of the organization and its competitive position in the market. Since the initial estimates of these parameters made during the early planning stages are usually not very accurate, they need to be adjusted to account for problems and challenges encountered during the program. Costs and time overruns are universally hated by the management. On the contrary, completion of the program before its planned end-date and under its budget are reasons for celebrations of the great accomplishments and deserve special recognition of the program teams. The revenues are received after the products are sold and customer satisfaction affects future sales.

REFERENCES

Blanchard, B. S. and W. J. Fabrycky. 2011. *Systems Engineering and Analysis*. Fifth Edition. Upper Saddle River, NJ: Prentice Hall.

Campanella, J. 1990. *Principles of Quality Costs*. Second Edition. Milwaukee, WI: ASQC Quality Press.

Hussain, T. and S. Randive, 2011. *Defining a Low Cost Vehicle for the U.S. Market*. Dearborn, MI: Institute for Advanced Vehicle Systems, College of Engineering and Computer Science, the University of Michigan-Dearborn. www.engin.umd.umich.edu/IAVS/books/A_Low_Cost_Vehicle_Concept_for_the_U.S._Market.pdf (accessed August 3, 2012).

24 Special Driver and User Populations

AN OVERVIEW ON USERS AND THEIR NEEDS

The purpose of this chapter is to provide the reader with insight into different populations of vehicle users and the differences within the populations. It must be realized that each population has some unique set of characteristics and needs that must be considered in designing an automotive product for its intended market segment.

The population of vehicle users can be distinguished by considering factors related to: (a) vehicle type and body style (e.g., owners of certain types of vehicles have unique needs); (b) geographic locations of the markets (specific countries) in which the vehicle will be sold; (c) type of uses (e.g., personal, family or work/commercial related uses); (d) level of luxury based on income, image and technology expectations of the customers and users; (e) educational and technical/ professional background of the users; (f) gender-specific use issues and male-to-female ratio of users; (g) user age and life stage related characteristics and needs; and (h) physical abilities and disabilities of the users.

The ergonomics engineers assigned to a vehicle development program need to thoroughly understand the population and sub-populations of the users of the planned vehicle. To understand issues associated in each of the above categories, the ergonomics engineers need to search for information available within the company, from sources or databases such as lessons learned from previous vehicle development programs, market research, customer feedback and complaints, and internal and external requirements on the vehicle. In situations where the vehicle type being designed is very different from the existing vehicles, the engineers need to gather information by visiting potential customers, observing how they use their vehicles and asking them questions related to their needs and preferences.

For example, when the author was asked to provide ergonomics support to a heavy truck design program, he not only learned to drive the heavy trucks with their long trailers, but performed many other tasks that the drivers routinely do, such as docking and undocking the trailer, hooking and unhooking air hoses and cables, filling fuel, checking fluids, and so forth. He also took long trips in heavy trucks and observed what the drivers did and asked them many questions about their equipment usage situations and problems. The experience helped in designing survey questionnaires

DOI: 10.1201/9781003485605-11

and in interviewing the truck drivers who participated in market research clinics and in planning special ergonomics studies.

Some special ergonomic studies conducted to study the heavy truck driver issues involved (a) in asking truck drivers to create their own full-size instrument panel mock-up using Styrofoam blocks, Velcro strips and a box full of controls and displays; (b) asking the drivers to enter and exit from the trucks equipped with different step heights and grab handles and video recording their hand and foot placements and movements; and (c) interviewing drivers on special situations involving field-of-view problems (e.g., visibility of low sports cars, pedestrians, bicyclists hidden behind long truck hoods, and high seatbelt lines).

UNDERSTANDING USERS: ISSUES AND CONSIDERATIONS

Traditionally, ergonomics engineers examine the anthropometric characteristics of users to make sure that they can be accommodated in the vehicle space. However, to truly meet the needs of customers and users, ergonomics engineers also need to gather information on many aspects that affect their usage experience. The issues associated with the usage experience are not just based on the demographic and educational background of the user population but also on the users' needs (e.g., loading/unloading items and fastening kids into child seats) and expectations on how various features of the vehicle should work. The factors to be considered here are grouped according to major influencing variables and described below.

VEHICLE TYPES AND BODY STYLES

The type of vehicle and body style that a user will select primarily depends upon the user's desires (e.g., sportiness, styling, performance, fuel economy, ride comfort, affordability) and transportation needs (i.e., carrying capacity in terms of number of passengers and load – sizes and weights). The vehicle type can be classified as: passenger car, SUV, crossover, truck (light, medium and heavy), van, or multi-passenger bus. The body style can be discriminated by features such as the SgRP height from the ground, number of occupants and seating configuration, number of doors and types of doors (e.g., hinged doors, traditional front hinged versus suicide doors, sliding doors, gull wing doors, hatchback, liftgate, tailgate, dual swing gates, etc.), presence or absence of the hood (e.g., a truck with a long hood versus a cab-over design), and cargo area (e.g., flat bed/open, closed, size of cargo box/bed).

In the United States, the male-to-female ratio of drivers in passenger cars is about half and half. The smaller and economy passenger cars, typically, have a higher percentage of female drivers (about 60%), whereas the larger and more expensive passenger cars have higher percentages of male drivers. US truck products, overall, have higher proportions of males as drivers. The proportion of males in the populations of the pickup truck drivers, the medium truck drivers, and heavy truck drivers are approximately, 75–85%, 85–90% and over 95%, respectively. It should be noted that the population of drivers (i.e., actual users) of automotive products could be different from the population of owners or purchasers of the vehicles. This is especially important when considering truck products. The truck owners (or purchasers) are generally different from the passenger car users because higher percentages of trucks are owned by businesses or government agencies.

In the past pickup trucks were mainly utility vehicles or "work horses" with less concern for driver comfort or ease of use. Over the past 20–25 years, that trend has been changing as many pickup trucks are owned and used by individuals who expect the luxury features and comfort of passenger cars, and still want the durability and utility of the trucks. Many models of light trucks strive for the same comfort, efficiency, drive-ability, and safety as the other passenger vehicle segments. At the same time, light trucks must be designed to meet their primary function of utility and meet the needs that other passenger vehicles are not well suited for usages such as hauling, off-road driving, and towing.

There are also vehicles that are truck-based and serve specialized functions such as recreational vehicles, vans with different seating arrangements, delivery trucks, ambulances, buses, garbage trucks, fire trucks, cement trucks, and so forth. They are typically sold by the vehicle manufacturers as incomplete vehicles (having fully drivable rolling chassis with or without a full cab); and other body builders build bodies with specialized features. The ergonomic considerations of such vehicles are, thus, handled by ergonomics engineers working for the body builders, outside consultants or the vehicle acquiring organizations (e.g., transportation companies).

MARKET SEGMENTS

The market segments for passenger cars are typically classified by vehicle size (e.g., subcompact, compact, intermediate, full-size), body style, price (e.g., economy, entry luxury, luxury, ultra luxury), and countries where the vehicles will be sold. The needs in the different market segments are driven by a number of considerations.

Some issues underlying the considerations are:

1) Anthropometric Differences: For example, Asian populations are shorter and require pedals to be located more rearward as compared to the vehicles sold in United States and Europe.

2) Country-related differences: The narrower roads, traffic congestion, higher fuel prices, lack of available parking spaces, and parking costs have an effect on types and sizes of vehicles used in different countries.

3) Economic differences: For example, manual transmission usages are higher in Europe and Asia (primarily because of the higher fuel prices as compared to the United States).

4) Changing trends and expectations: In recent years, the customers have high expectations of many vehicle features across all market segments. Small car buyers also are loading up on high-tech options, such as the Ford's hands-free Sync multimedia system, which is driving up the transaction prices (Pope, 2009).

5) Electric vehicle sales have been slowly increasing over the past few years. The higher torque provided by the electric motors at low speeds, low emissions

and reduced operating costs in comparison to the internal combustion engine equipped vehicles are some of the key advantages of the electric vehicles. Further cost reductions in battery technologies and increases in charging stations should increase sales of electric vehicles.

FEMALE DRIVERS

Female drivers have different needs and problems during vehicle usage than male drivers. On average, females are shorter than males (see Figure 24.1). They sit more forward and closer to the pedals and the steering wheel (if the pedals and the steering wheel are non-adjustable). The shorter females may also find insufficient clearance spaces for their knees (under the steering column and rearward of the knee bolster located below the instrument panel). Some females with shorter lower leg lengths may have difficulties supporting their heels on the floor and reaching the pedals (especially in vehicles with taller seat heights [H30 dimension] such as vans). The seated eye-heights of females are lower than for males. Thus, their view of the road over the steering wheel, instrument panel and the beltline would be somewhat limited as compared to taller male drivers. Shorter females will also find taller vehicles more difficult to get in and out of as compared to the taller males. They may find it difficult to reach open lift gates in vans and taller SUVs. Most females are less strong in

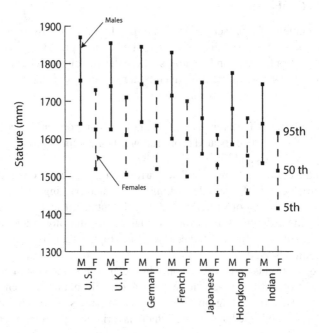

FIGURE 24.1 Comparison of 5th, 50th and 95th percentile values of stature of male and female adults from seven countries.

Source: Plotted from data provided in Pheasant and Haslegrave, 2006.

muscular strength than males. Thus, lighter forces in operating pedals, hand brakes, liftgates, hoods, folding seats, etc. must be considered. Many females have longer fingernails (95th percentile fingernail length among U.S. office workers is about 12 mm). Thus, door handles and controls need larger clearances to accommodate longer fingernails. Females carry purses and they need purse storage room inside or near the center console. Pregnant drivers would find adjustable (tilt and telescopic) steering column useful in obtaining larger stomach clearance.

OLDER DRIVERS

The population of the United States is aging at a rapid pace (see Figure 24.2). By 2030, it is predicted that about 71.4 million or 20% of the U.S. population will be older (age 65 and over) (U.S. Census Bureau, 2010). The population of older (age 65 and over) persons in the United States is estimated to increase from about 13% (40 million) in 2010 to 21% (87 million) in 2050 (refer to Figure 24.2). By the year 2020, drivers aged 65 years and older accounted for about 16.2 percent of the driving population. The fastest-growing group is the old-old category, which includes drivers aged 75 and over. Research to improve the quality of life for the elderly has been on-going for over 40 years or so with many studies appearing in journals and books (Fisk, Rogers, Charness et al., 2004; Charness and Parks, 2001).

As humans age, some anthropometric dimensions also change. For example, average stature of males as well as females decreases by about 12 to 15 mm (0.5 to 0.6 inches) per decade on average after about the age 30. Muscle strength decreases with increase in age. Thus, the older driver will find entering and exiting from vehicles more difficult. Older females, particularly, find entering and exiting from taller vehicles such as the full-size pickups and SUVs more difficult (Bodenmiller et al., 2002). The incorporation of running boards and assist straps and handles should

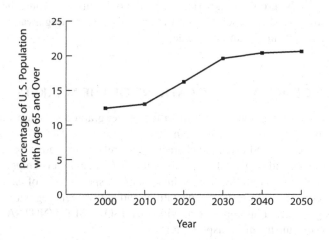

FIGURE 24.2   U.S. population projection of people age 65 and over.

Source: U.S. Census Bureau, 2010.

be considered to improve ease during entry/exit from taller vehicles. Tasks requiring higher physical efforts such as lifting rear gates and hoods, folding seats, loading and unloading heavier objects, etc. are also more difficult for older drivers. Due to higher incidences of arthritis and resulting in reduced range of hand, leg and body motions, the older driver will find the following tasks more difficult: reaching and pulling the seat belt buckle from its B-pillar anchor location, turning the ignition key, operating thumb activated gear shifters, unlatching seat belts, operating inside and outside door opening handles, operating hand brakes, and turning head while using a side view mirror.

Normal aging is associated with declining visual, attentional, cognitive, and physical abilities. Medical conditions also can accelerate deterioration in driving performance among older drivers. Due to degradation in visual functions with age and developing cataracts, the older driver would require larger letter sizes and higher contrast labels and graphics in visual displays (Bhise and Hamoudeh, 2004. See Volume 1, Figures 14.3 and 14.5). The older drivers also take longer times to detect, search and make decisions, especially in choice reaction time situations. Their visibility distances to visual targets at night are considerably shorter than those of younger drivers and the older drivers are also more affected by the effects of discomfort and disability glare (Bhise, Matle and Farber, 1989). As the drivers age, their hearing abilities likewise decrease, especially in the higher frequencies. Thus, frequencies above 2000 Hz in auditory warning signals should be avoided. The older drivers are also less willing to use new technologies, and they drive less during inclement weather and at nighttime.

In an effort to design vehicles for older buyers, Ford engineers developed a "Third Age Suit," which helps simulate specific problems associated with age, such as reduced sense of touch, vision, body movements and muscular strength (Ford Motor Company, 2002). Since wearing the suit adds about 30 to 40 years to the user's age, after wearing it while using the vehicle the younger vehicle designers and engineers quickly realized the problems experienced by the older drivers. Thus, the Third Age suit can be considered as a useful tool for ergonomists in creating awareness of older driver problems within the automotive design community and also to evaluate vehicle designs.

EFFECT OF GEOGRAPHIC LOCATIONS OF THE MARKETS

As pointed out earlier, the drivers in different geographic locations of different markets in different countries have some unique problems. The driver needs differ due to differences in conditions and characteristics related to roads (e.g., road geometry, road surface conditions, traffic speeds), climate (e.g., extreme temperatures, rainfall, snow, sand storms), traffic (e.g., vehicle density, speeds, mix of cars-to-trucks), languages, economy, culture, user expectations and government regulations on motor vehicle designs (e.g., requirements provided by ISO, USDOT/NHTSA, European/ECE, Japanese Ministry of Transport, etc).

The differences in anthropometric and strength characteristics between drivers from Western and Asian countries must be considered (Pheasant and Haselgrave,

2006). Natu and Bhise (2005) found differences in the needs of drivers from the United States, India and China while creating a low mass vehicle. Some specific issues that need consideration are (a) driver location in the vehicle (e.g., the left-hand drive versus the right-hand drive. Note that some countries, namely, the UK, Australia, India, and Japan drive on the left side of the road and thus use right-hand drive vehicles.), (b) driver positioning with respect to the pedals (Note: Longitudinal locations of the pedals depend upon driver stature, leg lengths and the presence or absence of the clutch pedal.), and (c) distributions of driver characteristics such as age, males-to-females mix, languages used, habits, and direction of motion stereotypes in operating controls and reading displays.

The differences in fuel prices, availability of parking spaces, traffic congestion and roads (geometric designs of the roadways, e.g., road widths and curvatures) between the US, European and Asian countries have affected the size of vehicles prevalent in different countries. The Asian and European countries have vehicles with higher percentage of manual transmission vehicles and the percentages are changing with advances in technology and changes in global economy. The effects of recent changes in economic and energy situations are also changing global market trends towards smaller and more fuel efficient electric vehicles. Further, to reduce costs, many manufacturers are sharing components and vehicle platforms and also creating global vehicles. The ergonomics engineers have the challenge to determine the areas where changes in design are necessary to accommodate drivers in different populations.

DRIVERS WITH DISABILITIES AND FUNCTIONAL LIMITATIONS

With the increase in the population of older drivers, the demand for increased mobility among the disabled and persons with various functional limitations will continue to increase. To maintain the mobility of drivers who have various disabilities, a number of vehicle modifications are available. The vehicle manufacturers need to consider the requirements to accommodate many common disabilities during vehicle designs so that the vehicles can be more easily modified to accommodate changes such as wheelchair access (modifications to doors/seats, addition of ramps or chair lifts), ease of grasping controls, and ease of reading displays.

With greater use of electronics, wireless, drive-by-wire technologies, and power assisted features it will be easier in the future to modify vehicles with fewer compromises for space and occupant protection issues. For example, an optimized joystick or lever steering control may offer mobility to people with one limb who have neither the range of movement nor the dexterity to operate a multiple-turn steering wheel. However, the alternative controls can be difficult to learn and use effectively can and impose significantly different workloads for different individuals.

Better fastening systems for wheelchairs, and restraint systems for occupants in wheelchairs in vehicles are also major needs. The restraint systems for occupants in wheelchairs are still difficult to use. An ideal system will not require special wheelchair hardware or significant effort to fasten. With further advances in rehabilitation medicine and adaptive technologies, many disabled people who never considered the driving possibility may be able to drive in the future.

ISSUES IN DESIGNING GLOBAL VEHICLES

In order to reduce vehicle development costs, all manufacturers are considering approaches such as developing vehicles with common platforms, common parts and common overall designs in creating vehicles that can be introduced in many countries. However, the ergonomics engineers will still face many unique problems in determining what can be commonized and where different designs are needed to meet the various needs in different countries.

There are some unique design problems in creating a truly common global vehicle design:

a) Differences between United States, European and Japanese standards: Many design procedures and requirements still are different for different countries (e.g., vehicle lighting and field of view standards) in spite of the many actions undertaken to harmonize standards.

b) Left-hand drive (LHD) versus right-hand drive (RHD): Left-hand drive and right-hand drive vehicles cannot simply be mirroring images of each other. For example, in all vehicles the accelerator pedal is always operated by the right foot. On the other hand, in many vehicles sold in European countries, the turn signal function is not always placed on the outboard stalk. Cost and commonality requirements also dictate that certain items, such as instrument clusters and center stack devices are not made or installed as mirror-imaged components for LHD and RHD vehicles.

c) Vehicle size (passenger cars and truck): The differences in lane widths, road curvatures, traffic congestion, parking spaces, fuel costs, and so forth dictate different sized vehicles in different countries.

d) Type of driving, trips and distance traveled per driver: Availability of alternate modes of travel (e.g., trains and buses) also affects vehicle uses and driving needs in different countries.

e) Differences due to user expectations and stereotypes: e.g., "up for ON" direction-of-motion stereotype for toggle switches in the United States versus "down for ON" in the UK and other Commonwealth countries.

f) Differences due to languages in user interface designs: for example, labels (words used for identification and setting labels, units used in displays [km/h versus MPH; gallons versus liters], and voice commands and voice recognition systems must be different to satisfy drivers with different languages.

FUTURING

"Futuring" is a term used by some market researchers in the automotive industry to predict future trends in vehicle designs. Reducing the planning horizon for vehicle development helps to reduce serious errors in determining customer needs for a future vehicle as accuracy in predicting the near-term future is generally better than in predicting the far-term future. Ergonomics engineers need to account for future changes in user populations, their expectations and changes due to design trends and technological advances. For example, the introduction of electric vehicles requires additional driver interfaces related to recharging needs, ease during recharging and reducing the "range anxiety" or the uncertainty related to completing the return trips (Bhise, Dandekar, Gupta and Sharma, 2010).

CONCLUDING COMMENTS

Vehicle development teams must ensure that they understand the population of users of the planned vehicle. This need would require the team to foresee special populations, their needs and problems. Designing the new vehicle to fit the special populations and their needs will extend the market penetration and customer satisfaction with the vehicle. Ergonomics engineers must make extra efforts to study the special populations and ensure that the design tools (e.g., seat track design for females and Asian populations with shorter stature and leg lengths) and design guidelines used during the design process meets the needs of the intended special populations.

REFERENCES

Bhise, V., H. Dandekar, A. Gupta and U. Sharma. 2010. *Development of a Driver Interface Concept for Efficient Electric Vehicle Usage.* SAE paper no. 2010-01-1040. Paper presented at the 2010 SAE World Congress, Detroit, MI.

Bhise, V. D. and R. Hammoudeh. 2004. A PC Based Model for Prediction of Visibility and Legibility for a Human Factors Engineer's Toolbox. *Proceedings of the Human Factors and Ergonomics Society 48th Annual Meeting*, New Orleans, LA.

Bhise, V. D., C. C., Matle and E. I. Farber. 1989. *Predicting Effects of Driver Age on Visual Performance in Night Driving.* SAE paper no. 890873. Presented at the 1988 SAE Passenger Car Meeting, Dearborn, MI.

Bodenmiller, F., J. Hart and V. Bhise. 2002. *Effect of Vehicle Body Style on Vehicle Entry/Exit Performance and Preferences of Older and Younger Drivers.* SAE Paper no. 2002-01-00911. Presented at the SAE International Congress in Detroit, MI.

Charness, N. and D. C. Parks. 2001. *Communication, Technology and Aging: Opportunities and Challenges for the Future.* New York, NY: Springer Publishing.

Fisk, A. D., W. A. Rogers, N. Charness, S. Czaja and J. Sharit. 2004. *Designing for Older Adults: Principles and Creative Human Factors Approaches.* Boca Rotan, FL: CRC Press.

Ford Motor Company. 2002. Ford Drives a Mile in an Older Person's Suit. www.design.ncsu.edu/cud/projserv_ps/projects/case_studies/ford.htm (accessed: August 10, 2005)

Natu, M. and V. Bhise. 2005. *Development of Specification for UM-D's Low Mass Vehicle for China, India and the United States.* SAE Paper no. 2005-01-1027. Presented at the 2005 SAE World Congress, Detroit, MI.

Pheasant, S. and C. M. Haslegrave. 2006. *BODYSPACE: Anthropometry, Ergonomics and the Design of Work.* Third Edition. London: CRC press, Taylor and Francis Group.

Pope, B. 2009. Ford Pulling Itself Up by Its Bootstraps. *Ward's AutoWorld*, December 1, 2009. http://wardsautoworld.com/ar/auto_ford_pulling_itself/ (accessed: March 1, 2010)

U.S. Census Bureau. 2010. Population Projections. Projected Population of the United States, by Age and Sex: 2000 to 2050 Table 2a. www.census.gov/programs-surveys/popproj.html (accessed: March 17, 2024)

25 Future Research and New Technology Issues

INTRODUCTION

Over the past 38 years, the applications of ergonomics in vehicle designs have improved user comfort, convenience and safety substantially. Compare any 1985 model year vehicle with its 2023 model year vehicle. You will find that almost every item that the driver interfaces in these vehicles has changed substantially. For example, many 1985 model year vehicles had radios with two rotary knobs, five push button presets and a sliding pointer display to set a radio station. The climate controls were mechanical cable-operated sliding or rotary controls. Now, the 2023 model vehicles have radios with many features such as AM, FM, satellite stations, CD, and a USB or Bluetooth interface to connect a number of digital media. In many vehicles, all the radio functions can be operated by voice controls. The touch control displays are in center stacks of most vehicles and allow selection of many features and applications such as navigation, audio, vehicle settings, and cellphone directory. The newer climate controls have features such as precise temperature settings for the driver and the passenger, automatic climate control, controls for the rear occupants, ability to view outside temperature, and so forth. The climate controls in many 2023 model year vehicles have redundant controls that are accessible from the center touch screen as well as the use of dedicated controls (due to safety concerns to find certain functions, such as the windshield defrost, quickly when needed). In many newer vehicles, the phone, radio and climate controls can be operated by redundant steering wheel mounted controls that have lighted labels for night legibility, and various electronic displays can be selected (from preconfigured layouts) and set with these controls. There are at least 50 or more individual controls in front of the driver in newer vehicles (see Figure 25.1) with well-illuminated labels. Use of many colors, functional grouping, selection of preferred displays and decluttering of displays is possible with reconfigurable displays. The change also has led to extending feature content to many economy vehicles.

The above-described situation, thus, illustrates that with the rapid advances in many technologies, future needs in implementing ergonomic solutions are even greater. The needs are both in developing improved methods to design and evaluate the products, and in developing products with improved functionality, features, safety and convenience.

DOI: 10.1201/9781003485605-12

FIGURE 25.1   Instrument panel controls and displays in a late model year vehicle.

The purpose of this chapter is to cover future research issues related to improving applications of vehicle ergonomics and implementation of new technologies in designing future vehicles. The chapter addresses the future of automotive ergonomics in terms of issues related to details such as configurations, types of interfaces, operability, distractions and driver workload and what ergonomic data and tools are needed to design future vehicles.

ERGONOMIC NEEDS IN DESIGNING VEHICLES

In order to understand the ergonomic needs in designing new vehicles for the future, let us first review the basic driver needs, which can be considered to fall in the following broad areas:

1. Providing greater levels of mobility, comfort and convenience to vehicle users
2. Improving safety (freedom from injuries through crash avoidance and crash protection)
3. Improving productivity and reducing costs (e.g., efficient utilization of time, reduced cost/mile, fuel economy)
4. Providing a pleasing environment (e.g., choice in entertainment features for the drivers and passengers, well-crafted and more spacious vehicle interiors)
5. Creating "Fun to Drive" vehicles (i.e., pleasing sensory perceptions during vehicle usages)

Considering the above needs along with the Kano Model of Quality and the Ring Model of Desirability (covered in Volume 1, Chapter 12), the advances in technologies should offer features that would not be just expected by the customers, but they should also increase comfort/convenience, create pleasing perceptions and delight customers. Now, let us jump directly into the current automotive scene and define the needs and expectations of the users of passenger vehicles that can be introduced in the near future, that is, within the next five years. The following section presents a summary of customer needs in designing passenger vehicles.

VEHICLE FEATURES IN THE NEAR FUTURE

The following features are expected in future automotive products.

1. High-tech cars – safer, more enjoyable, more comfortable and smarter – that is, can configure functions to individual needs, anticipate unsafe situations and warn the driver or perform certain functions to assist the driver
2. Smaller cars – more emphasis on styling, high fuel economy and reduced emissions
3. Crossover–type vehicles that can fill the needs met by SUV and minivan styling, but with higher fuel economy and reduced emissions
4. Better fitting interiors – greater occupant accommodation, for example, comfortable driver space that accommodates higher percentage of users by means of adjustable seats, pedals, and tilt/telescoping steering columns
5. More storage spaces for items to be stored in the vehicles
6. Crash avoidance systems (e.g., improved braking, handling, stability, driver assistance systems [e.g., drowsy driver or alertness monitoring and warning systems], smart lighting and visibility systems)
7. Passive safety systems (e.g., smart airbags, pre-tensioning belts, side curtains)
8. Comfort systems (e.g., automatic climate-control systems for each seating location, heated/cooled seats)
9. Convenience features (e.g., more personalized memory settings of favorite radio stations, seats, mirrors, blind spot sensors, air-conditioning, display options and menus, reconfigurable seating, parking aids, and cruise controls)
10. State-of-the-art entertainment systems (e.g., radio with AM/FM, CD, satellite radio, plug-in or wireless capabilities for aftermarket entertainment and gaming systems for the rear seat occupants)
11. Information and communication systems (e.g., Bluetooth or other wireless connectivity to cell phones and other handheld devices for a variety of functions, internet search and e-mailing, navigation systems with real time traffic, file/data management and storage, vehicle diagnostic system)

The challenges of ergonomics engineers are to work with other team members in meeting the above customer needs by implementing technologically feasible features and simultaneously meet the corporate business needs (primarily keeping the costs and timings under control).

FUTURE RESEARCH NEEDS AND CHALLENGES

ENABLING TECHNOLOGIES

Telematics is perhaps the key set of enabling technologies for new features in the driver information interface area. Telematics can be defined as a discipline which has emerged from the coming together of electronics, communications and information technologies. The French word télématique was coined in the 1970s to denote the combination of télécommunications and informatique "computing". Thus, "Automotive Telematics" is an inter-disciplinary applied field that deals with the development of devices for communication of information within the vehicle and other external sources through integrated applications of technologies such as wireless data transfer, sensors, micro-processors, databases, displays, lighting, actuators, controls, and global positioning system (GPS) for convenience of vehicle users. Automotive telematics, thus, facilitates the creation of a "Digital Car".

Many features or functions in future in-vehicle devices can be created by combining the above capabilities. For example, combining the GPS capability and databases on locations of gas stations, their service capabilities and prices, a display screen can provide the driver information on upcoming gas stations on the navigation screen along with their special service capabilities.

Thus, the "Telematics" provides a platform for services and content that could offer value to the vehicle users through combinations of the following:

a) Two-way wireless communication to external and within vehicle sources, for example, Bluetooth phones, V2X capabilities, that is, vehicle to X where X stands for drivers in other vehicles, infrastructure, home, other persons)
b) Use of databases on the internet and other data centers
c) Location technologies (GPS, cell phone)
d) Vision systems (with sensors, cameras and displays with superimposed processed information)
e) Voice technologies (voice recognition, voice controls and voice/auditory displays)
f) Reconfigurable driver interfaces (reconfigurable displays and multi-function controls)
g) Sensors, signal processors, actuators and control units

The specialized technologies contributing to advances in-vehicle features can be categorized as the following:

1. Driver Interface Technologies: Reconfigurable displays, touch displays, gesture controls, multi-function controls, voice technologies, and steering wheel mounted controls
2. Driver State Measurements and Assistance Systems: Drowsiness detectors, driver workload monitors, intoxicated (under the influence of alcohol and/or drugs) driver detection and crash avoidance warning systems
3. Vehicle Status Information and Diagnostics: for example, hybrid powertrain energy flow display, state of battery charge in an electric vehicle, and service alerts.
4. Longitudinal and Lateral Control Systems: for example, adaptive cruise controls (ACC), lane centering and control system, enhanced stability control (ESC) and enhanced braking systems

5. Ride and Thermal Comfort: adaptive suspension systems, advanced thermal comfort systems
6. Lighting Technologies: for example, smart headlamps, new technology signal lamps and interior lighting (LEDs, light piping, and electro-luminance)
7. Quality, Craftsmanship and Brand Sensory Perceptions: Selection and applications of interior materials with pleasing tactile feel and visual harmony, controls with pleasing tactile feel (e.g., crispness and smoothness felt during operation of switches), and interior lighting for display legibility, locating objects and pleasing visual and sound effects inside the vehicle

CURRENTLY AVAILABLE NEW TECHNOLOGY HARDWARE AND APPLICATIONS

Many technology applications presently in use or close to near-term introductions are briefly described below.

1. Digital LCD Instrument Clusters: LCD displays allow for crisper high-contrast graphics, with colors and reconfigurability (i.e., depending upon the driving situation, different information such as gages, status of different vehicle or trip parameters or camera views, can be presented).
2. 3-D Displays: These displays allow presentation of information that can appear to be located at different viewing distances. The apparent depth of objects in the displays can be used for functional grouping of displayed items.
3. Touch Screens and Multi-touch Technology: The user can activate different display modes by touching different control areas (e.g., touch buttons) on the screen. Multi-touch technology will allow recognition of multiple touch areas and movements of the touching fingers into preprogrammed control actions (e.g., selecting certain screens or functions).
4. Steering Wheel Mounted Controls: Push buttons, rockers or rotary controls mounted on the steering wheel spokes or hub areas allow the drivers to select and operate the controls without moving their hands away from the steering wheel.
5. Multi-function Controls: A combination control that involves multiple controls grouped together or a single control with multiple functions assigned to its different activation movements, for example, pushing, pulling, rotating and/or moving in different directions like a joystick with different haptics feedbacks for different switch modes.
6. Projected Displays: Projectors mounted behind one or more screens can show (back projected) displays. The screen can include touch areas as well as programmable (soft) labels for some hard controls.
7. Head-up Display: A display projected on a partially reflective glazing surface to form an image focused on a farther distance and located such that it can be viewed by the driver without turning his head down (i.e., maintaining "head-up" orientation).

8. Dual-view Screen: A display screen which when viewed from different directions (e.g., from the driver's eye location and a front passenger eye location) provides different images. For example, a center stack mounted dual display screen can provide the driver a navigation display, whereas the front passenger can view a video entertainment channel.

9. Flexible Displays: Displays mounted on a flexible substrate (like a paper). The flexible displays can be formed and applied (or wrapped) on any complex-shaped surface (non-flat). Also, the flexible display can be folded or rolled (like a window-shade) for storage purposes.

10. Navigation Map Display Formats: The map can be presented in different formats (e.g., traditional map, map oriented to conform to the forward direction of travel, or a bird's-eye-view perspective with roadside objects or landmarks).

11. Rear Passenger Entertainment Systems: The displays for the rear entertainment system can be roof-mounted flat panel displays, separate flat panel screens for each designated sitting position, or handheld/portable devices that can present TV programs or outputs of DVDs or game players.

12. Satellite Radio: The services offer coast-to-coast, mostly commercial-free radio with a large variety (from rap to opera, sports to children's programming). Currently, XM and Sirius each offer many channels.

13. OnStar System (from General Motors): This system involves a control panel with push buttons mounted inside the vehicle within the driver's reach and vision zones. The OnStar center is contacted with a push button or automatically in case of an accident (through a cellular call). The OnStar center's advisor can contact the driver to provide assistance under a number of situations such as: when an air bag is deployed, stolen vehicle tracking (help police with vehicle location), remote door unlock, driving directions, roadside assistance, remote diagnostics, personal calling (hands-free call with voice controls), remote horn and light activation (flashes lights to find the car), accident assist (e.g., it calls police), ride assist (e.g., it calls a cab), online concierge (e.g., locates a restaurant), and a personal concierge (e.g., helps getting tickets). Several other vehicle manufacturers offer systems with similar features.

14. SYNC (from Ford Motor Company): The SYNC is an in-car connectivity system that allows front seat occupants to operate the most popular MP3 players, Bluetooth-enabled phones and USB drives with simple voice commands. The SYNC features include turn-by-turn directions, 911 assist, vehicle health reports, news, sports and weather, real-time traffic, and business search.

15. Night Vision System: This system helps the driver detect objects before they are visible under headlamp illumination. It uses an infrared camera to view objects beyond the driver's visible areas with headlamps and presents the view to the driver on an instrument panel-mounted screen or through a head-up display.

16. Lane Departure Warning System: The system provides warning alerts when the vehicle leaves its lane.

17. Lane Centering System: This system automatically controls the vehicle path within the lane.

18. Forward Collision Warning System: The system uses a combination of satellites, radar and/or electronic sensors to determine if a driver is too quickly approaching a slower or stopped vehicle (or a fixed object in the vehicle's path). It alerts the driver with a series of beeps and visual signals on an interior display. The signal warns the driver to brake or make an evasive maneuver.

19. Front Adaptive Cruise Control: This cruise control adjusts the speed of the vehicle to follow a leading vehicle at a preset distance (headway) and maintain speed at or below a preset speed.

20. Rear and Side Vision Aids: These systems provide warning signals if another vehicle or an object (e.g., pedestrian) is within any of the driver's maneuvering or blind areas. Examples of such systems are a) rear object detection systems, b) rear and side camera systems, and c) blind area (blind-spot) detection systems.

21. 360 degree camera: It provides a plan view of the vehicle, showing and alerting to objects close to the vehicle or moving toward a collision course with the vehicle.

22. Voice Controls/Recognition Systems: The voice systems can a) recognize spoken words and select settings of control functions, b) reduce driver's workload by eliminating eye and hand movements to controls, that is, provide for "hands-free" operation, c) present long voice messages from the vehicle, and d) use only selected vocabulary for a faster response. (Some problems with such systems are slower response time and poor voice recognition accuracy, especially under a noisy moving-vehicle environment.)

23. Text-to-Speech Conversion Systems: The driver can select different languages as well as voice, e.g., male/female. Several text-to-speech software implementations can present voice in different languages and dialects such as US English, Continental French, Latin American Spanish, UK English, German, Japanese, Brazilian Portuguese, Americas Spanish, Australian English, and Canadian French.

24. Driver state monitoring systems: These systems can detect the driver's state of alertness by monitoring the driver's control actions (e.g., steering wheel movements), eye closures, or physiological state.

25. Memory Seats: The seat track location and settings of seat cushion and seatback angles, contouring and padding can be memorized and readjusted for entry/egress convenience and seating comfort.

26. Heated and Cooled Seats: These seats can improve the occupant's thermal comfort by adjusting temperatures of seat cushion and seat back.

27. Smart Headlamps: These LED lamps can provide more light flux at reduced wattage (as compared to the older tungsten filament lamps), and thus, can provide beam patterns with improved night visibility. The beam pattern also can be modified depending upon driving environment such as curves and vehicle speed.

28. Rain Sensor Wipers: This wiper control system can reduce driver involvement in starting, stopping or resetting of wiper speeds. The system sensor senses rainfall intensity and adjusts wiper speeds for improved visibility.
29. Remote Tire Pressure Sensing: This system provides warning messages when the air pressure in any of the tires is below a preset pressure level.
30. Advanced Climate Controls: This system provides improved thermal comfort by adjusting temperature, air flow location, direction, and flow rates by measurements from a number of sensors (e.g., temperature and radiant heat sensors).
31. Bluetooth Communications: This is a technology that allows users to wirelessly connect electronic devices (e.g., iPhones using CarPlay) to their cars and perform tasks like making hands-free calls without having to physically plug the devices into the vehicle.
32. Data Transfers: These systems will allow transfer of data wirelessly between the vehicle and other locations such as a) other vehicles, b) roadside transmitters and receivers, c) dealers and vehicle manufacturers, d) traffic and weather information centers, e) other service centers for locations of gas stations, restaurants, banks, stores, and so forth.
33. Vehicle Stability and Control systems: These systems improve directional stability of vehicles during maneuvers by selectively controlling wheel speeds and suspension characteristics (if adaptive suspensions are used).
34. Parking Aids: These systems will provide the driver with warning signals during parking or will perform the parallel parking task automatically.

A POSSIBLE TECHNOLOGY IMPLEMENTATION PLAN

A group of six graduate students in the author's 2010 Vehicle Ergonomics II class were given an assignment to study the technology trends and customer needs and determine feature contents in future vehicles. The students were asked to study an existing late-model light vehicle (a passenger car, a light truck or an SUV) and develop a technology implementation plan for the selected vehicle. Two time frames were considered for the assignment which included: a) a near-term model that could be introduced in the next 3 to 5 years, and b) a far term model that could be introduced in the next 5–10 years. Table 25.1 summarizes the outcomes of the assignment. Since the feature content will depend upon whether the selected vehicle is an "economy" or a "luxury" version and whether each feature be offered as "standard" or "optional" content, the recommendations for the future features were classified as: "SE" (standard economy), "SO" (standard optional), "SL" (standard luxury), "OE" (optional economy) and "OL" (optional luxury). Table 25.1 presents 19 features considered in the assignment and their allocation to the near and far term vehicle models using the above classification. The table also provides important ergonomic issues, ergonomic advantages and ergonomic disadvantages considered of each of the 19 features.

TABLE 25.1

Technology Implementation and Ergonomic Considerations

No.	Product Feature or System	Implementation Time Near Term (3–5 yrs)	Far Term (5–10 yrs)	Ergonomic Issues	Ergonomic Advantages	Ergonomic Disadvantages
1	Reconfigurable instrument cluster	SL	OE, SL	Display content, choice of predefined option groupings to customizable displays	Designed for personal preferences, reduced glance durations	Complexity due to more choices
2	Touch screen in the center stack	OE, SL	OE, SL	Legibility, comprehension, visual clutter, feedback on completed action	Simple pointing finger movement, contact grasp, additional coded feature possibility from multi-touch operations	Driver workload, menu structure, pointing errors due to vehicle movements
3	Head-up display	OL	OE, SL	Location, content, luminance, masking of other visual cues and possible distraction	Keep eyes on the roadway, reduced eye time to read displays, easier for older drivers	Masking part of forward field, distracting effects
4	Voice controls	OE, OL	OE, SL	Voice recognition error rates, recognition time	Reduces visual load	Voice recognition accuracy, recognition delays
5	Navigation system with real time traffic information and alternate route planning	OE, OL	OE, SL	Display content, menus, controls, driver workload	Reduced travel time and reduced driver fatigue	May add more steps and time in route selection. Increased driver workload

TABLE 25.1 (Continued)
Technology Implementation and Ergonomic Considerations

		Implementation Time				
No.	Product Feature or System	Near Term (3–5 yrs)	Far Term (5–10 yrs)	Ergonomic Issues	Ergonomic Advantages	Ergonomic Disadvantages
6	Data storage (hard disk or other memory)	OL	SL	Type and amount of data to be stored and retrieval procedure, menu structure, use of past/ historic data for personalization	Can reduce driver workload and uncertainty by displaying needed information, choice in accessing data according to individual preferences	Added complexity due to data storage and retrieval procedures, options and controls
7	Rear passenger entertainment system	OL	OL	Additional controls for the driver (decision to delegate control to the rear passenger), placement and size of displays	Reduced conversations and distractions from rear passengers	May increase driver distractions. Use of headphones. Increased noise
8	Rear climate control system	OE, SL	SL	Additional climate controls for the driver (decision to delegate control to rear passenger)	Increased passenger comfort	Added feature complexity
9	Forward collision warning system	OL	OL	Effectiveness of the warning signal, driver acceptance/ trust in the system	Reduced chances of collisions	May increase driver annoyance due to too many warning signals. Increased reliance can give false sense of security

(continued)

TABLE 25.1 (Continued)
Technology Implementation and Ergonomic Considerations

No.	Product Feature or System	Implementation Time		Ergonomic Issues	Ergonomic Advantages	Ergonomic Disadvantages
		Near Term (3–5 yrs)	Far Term (5–10 yrs)			
10	Lane departure warning system	OL	SL	Effectiveness of the warning signal, driver acceptance/ trust in the system	Reduce chances of lane departures and run-off the road collisions	May increase driver annoyance and workload due to too many warning signals
11	Blind area detection system	OE, OL	OE, SL	Warning signal effectiveness based on modality, location, intensity, frequency, driver acceptance/ trust in the system	Detection of targets in blind areas, reduced head movements and accidents related to obscured targets	High frequency of warning signals can lead to drivers disregarding the signals
12	Driver attention monitor	OL	OE, SL	Methods to measure alertness and their accuracy, driver's trust in the system	Alerts the driver before getting into hazardous situations	Increased reliance can give false sense of security
13	360 degree vision monitoring system		OL	Location and content of the display showing views from multiple cameras	Increased driver awareness of objects around the vehicle (Especially useful for commercial truck drivers)	Increased driver workload, adds another complex display
14	LEDs in rear signal system	SL	SE, SL	Effectiveness of the signal due to variations in aspect ratios of lighted lamp areas	Improved visibility, quicker rise time, long life, energy savings	Increased variability between rear signals of vehicles in the forward field

TABLE 25.1 (Continued)
Technology Implementation and Ergonomic Considerations

No.	Product Feature or System	Implementation Time		Ergonomic Issues	Ergonomic Advantages	Ergonomic Disadvantages
		Near Term (3–5 yrs)	Far Term (5–10 yrs)			
15	LED Headlamps	OL	OL	Variations in beam patterns between vehicles	Energy saving, can be used with smart head lighting concepts	uneven luminance distribution on the pavement
16	Smart headlamps	OL	OL	Beam selection and switching criteria and controls. Designing beam patterns. Prediction of expected and unexpected targets	Improved visibility and reduced glare effects on on-coming drivers	May reduce visibility if smart headlighting does not get correct road and vehicle state information
17	Park assist system	OL	OL	Increase in driver controls and learning to trust the system	Reduced driver workload, parking accuracy	Over reliance on the system and less attentive driver
18	Automatic user profile memory for personalization	OL	SL	High within subject variability can reduce effectiveness of the feature	Eliminate need to reset personal settings	Increased feature content can increase feature setting difficulty
19	Reclinable rear seats	OL	OL	Passenger accommodation and comfort, control location, range and speed of seat movements	Improve sitting comfort of rear occupants	Need to increase couple distance (longitudinal spacing between front and rear SgRPs)

The ergonomics engineers are often asked to participate in product planning activities and prepare a feature implementation plan as shown in Table 25.1 for technology planning and feature content planning for future automotive products (see Bhise [2017])..

QUESTIONS RELATED TO IMPLEMENTATION OF THE TECHNOLOGIES

The number of possible features that can be introduced in future vehicles will increase rapidly with advances in technologies. The ergonomics engineers will need to consider many issues during designing the systems and deciding whether the benefits that can be claimed would be indeed realized and the costs and disadvantages with their introduction can be minimized.

Many of the issues will require decisions related to details such as:

a) Locations of controls and displays
b) Selection of types of controls, displays and their layouts
c) Driver understanding and operation of the device
d) Compatibility with driver expectations, ease of use and distracting effects during operation
e) Effect of the new features on operation of other driver interfaces (e.g., it may obstruct an existing control, cause delays or interruptions in operation of other systems)
f) Priorities in displaying the information related to the new features with respect to information displayed from other features
g) Sharing or reconfiguration of functions within multi-function controls and displays and compatibility between different sensory modalities of displays (e.g., visual, auditory, tactile/touch)
h) Reducing complexity of the features/functions and the resulting driver workload

Driver distraction is one of the key topics of concern to government agencies as well as the automotive manufacturers. Some questions raised during the new technology implementations and driver distraction are:

a) How do in-vehicle technologies influence driver distractions? What are the effects of distraction on driving performance and safety? How do distractions from in-vehicle technologies differ from and compared to distractions due to other sources (e.g., talking to other passengers or eating while driving, using cell phones)?
b) What are the methodological challenges in measuring the influence of design features of the devices, their operation, and their impact on driver distraction and driving performance? (See Chapters 15 and 16).
c) What methods can be developed to relate measures of driver workload or distraction to the probability of an accident?
d) What actions can be taken by the government, the industry, and the consumers to minimize risks associated with different types of driver distraction?
e) What current and future research issues must be addressed to support actions to minimize driver distractions?

Naturalistic driving research (e.g., Klauer, et al., 2006; Sayer, Devonshire, and Flanagan, 2007) provides a unique means to understand real-world driver behavior and actual risks of crash and near-crash involvement. These data must be considered as the means to validate studies conducted in experimental, laboratory, or simulator settings. It is important to keep in mind that a short-duration study in a laboratory or simulator setting can only provide data on driver performance. Driver behavior can only be understood when observed over time in the real world.

Some basic questions with regard to the crash avoidance and warning systems include the following:

a) What is the proper time allowed for collision warning? Do multiple-stage warnings make sense for some types of crash hazards? What is an acceptable trade-off between false positive alarms and customer acceptance?
b) How much freedom should drivers have in selecting system functions for their own use?
c) How should multiple warnings for multiple crash threats be coded, combined, or sequenced?
d) How does new technology modify the driver's baseline levels of performance and behavior?

SPECIAL NEEDS OF GROWING MARKET SEGMENTS

PICKUP TRUCKS

1. Fuel economy and lower emissions: More efficient and lesser polluting engines are needed to meet more stringent future NHTSA/EPA requirements. Replacing large ICE engines with hybrid or electric powertrains would reduce fuel consumption and emissions. Incorporating larger batteries for the electrification of powertrains can partially off-set by incorporating light-weighting methods such as using light-weighting techniques (e.g., using light-weight materials such as aluminum, magnesium, carbon fibers and reconfiguring sections of different body, chassis and powertrain components,
2. Trailer Towing: Improving trailer-towing capacity and driver visibility to the rear and side of the vehicle
3. 4-wheel steer and air-suspensions: Four-wheel steering helps parking, maneuvering at low speeds (e.g., smaller turning radius), handling (e.g., crab-walking during lane changes). Incorporating a 4-wheel steering system in future pickups will significantly increase the maneuvering capabilities of the pickup truck as it will decrease the turning radius of the vehicle and also the effort to turn the vehicle. In this steering, the front and rear wheels turning angles are interconnected and are operated simultaneously. This steering system will provide better steering response, increases the cornering stability and the straight-line stability. This system will help the vehicle to park in tight corners, maneuver easily in curvy roads and provides low speed agility.

4. Air-suspensions: The air-suspensions allow ride height lowering for reduced aerodynamic drag and reducing foot lift-off to improve ease in entry/exit.
5. Additional vehicle sensors: 360 degrees vision cameras and sensors provide visibility of the trailer and traffic in adjacent and rear fields.
6. Convenience features: Rear doors in cabs (i.e., mid-gates), providing side gates, load box extenders, and tailgate and side mounted steps and handholds for access into the load box – for ease in loading/unloading items and in entry/egress.
7. Incorporating more driver assistance as standard features: Lane departure warning, lane centering assist, adaptive cruise control, and automatic high beams. SAE Automation level 3 or higher capabilities (see Chapter 18).
8. Large touch screen in center stack area of the instrument panels: A large touch screen with 12 inches diagonal length offers bigger areas for additional feature content for various infotainment capabilities. Allows design of larger-size and larger number of touch buttons in each screen and thus reduces flipping through many screens to select a desired control.

SUVs

1. Reconfigurable seats: Rear seats that can slide fore and aft to improve rear passenger legroom or increase rear storage space and flat-fold rear seats allow for more convenience in using the rear space.
2. In addition, many of the features listed above for pickups can be incorporated in the larger SUVs.

OTHER RESEARCH NEEDS

There are research needs both in the areas of development of improved products and improved methods for evaluating the products to ensure that future features will be perceived by the customers to be useful and value-added. The research needs are presented below:

1) Improved anthropometric databases: Larger anthropometric databases involving both additional measurements and measurements of different populations of drivers and users in different market segments are needed. Recent advances in laser scanning equipment such as that used in the CAESAR (Civilian American and European Surface Anthropometry Resource) project administered by the Society of Automotive Engineers (SAE) could be used for collecting whole body scan data (Reed et al., 1999).
2) Tools to accommodate a large percentage of user population with reduced costs and timings:
   a) Automated occupant packaging and visualization tools. The tools can perform basic occupant accommodation assessment (e.g., driver positioning, body clearances, location of major controls, and entry/egress) and conduct many specialized analyses such as field of view analyses for drivers

in different populations (Note that the present SAE standards, available in the United States, are based on US drivers).

b) Integrated digital car and digital manikins and visualization tools. CAD tools with manikin models (Digital Human Models), such as Jack/Jill, SAFEWORK, RAMSIS, SAMMIE, and the UM 3DSSP, are being currently used by different automotive designers to assist in the product development process (Chaffin 2001, 2007; Reed, Parkinson, and Chaffin, 2003; Bader, Allbeck, Lee et al., 2005; Human Solutions, 2010). Many of these tools are being updated to incorporate additional capabilities. Before using any of the models in the design process, the ergonomics engineer should conduct validation studies to determine if the postures assumed by the selected digital human model and their dimensional outputs indeed match closely with the postures and dimensions of real drivers under different actual usage postures and situations.

c) Programmable vehicle models (called PVMs). There are many different programmable vehicle models that are used in the early stages of vehicle packaging work. The models are physical bucks that can be configured very quickly in about (10–20s) by inputting selected package dimensions in a computer that controls many electric stepper-motors used to adjust the buck (e.g., Prefix Corporation, 2024). Such PVMs with improved capabilities to configure different sizes and types of vehicles and interfacing of the bucks with CAD systems would be useful in evaluating a number of vehicle design configurations. The PVMs have the potential of reducing time and costs in the early vehicle development process.

3) Understanding the needs of diverse vehicle users (including special population issues covered in Chapter 24). Systematic studies on different vehicle users' characteristics beyond just the anthropometric and biomechanical characteristics are needed. Future databases with information on characteristics related to driver vision, audition, cognition, touch feel perception, driver familiarity and expectations, preferences on interior materials and operation of controls for different market segments (and also to create global vehicles) are needed.

4) Driver interface prototyping tools. Computer-aided tools to generate operational prototypes of future in-vehicle devices are needed for evaluation of alternate design concepts in instrumented vehicles and/or driving simulators.

5) Driver interfaces for electric vehicles. As electric vehicles are gaining a larger market share, it is important that their designs be fine tuned to address customer concerns and needs. Due to limitations of the current battery technologies, recharging needs, availability electric vehicle servicing needs and safety concerns, electric vehicles may be perceived and used differently than the traditional vehicles powered by internal combustion engines. Ergonomics engineers, thus, need to address the following four problems: a) reducing driver uncertainties in using the electric vehicles – ensuring that the driver will get an accurate estimate of the state of charge (SOC) of the battery and how far the vehicle can be driven, b) minimizing of energy consumption by

providing information that can help the driver in adjusting or adapting his driving behavior, c) minimizing vehicle operating costs, and d) maintaining the vehicle in its best operating condition. A paper by Bhise, Dandekar, Gupta and Sharma (2010) provides additional information on the challenges in design and evaluation of driver interfaces for electric vehicles.

6) Methods to measure driver workload. Better methods that can measure driver workload, driver performance and driver distraction are needed (see Chapters 15 and 16 for more information). Further, criteria levels on a number of driver performance and workload measures are needed to determine acceptable (safe) and unacceptable (unsafe) levels of driver workload during different driving situations.

7) Driver warning systems. New methods to warn and alert drivers are needed to improve the effectiveness of future driver aids such as collision warning, driver alertness systems (e.g., drowsiness indicator).

8) Driver perception of quality and craftsmanship. Future research efforts need to be directed in understanding perception of quality and craftsmanship and their relationship to physical characteristics of materials used in interior components such as seats, instrument panel (including displays and controls), consoles, door trim panels and other trim parts.

CONCLUDING REMARKS

While a considerable number of research studies, design tools and methods exist for the evaluation of many ergonomic issues in vehicle design, more research studies and tools are needed to design future vehicles efficiently. Advances in new technologies and changing driver expectations related to capabilities of future vehicle features, perception of quality and craftsmanship will require ergonomics engineers to expand their knowledge and create new concepts, methods and tools.

REFERENCES

Badler, N., J., Allbeck, S.-J. Lee, R. Rabbitz, T. Broderick and K. Mulkern. 2005. New Behavioral Paradigms for Virtual Human Models. *SAE Trans. J. of Passenger Cars – Electronic and Electrical Systems*. Paper 2005-01-2689. Presented at the 2005 SAE Digital Human Modeling Conference, Iowa City, IA.

Bhise, V., H. Dandekar, A. Gupta and U. Sharma. 2010. *Development of a Driver Interface Concept for Efficient Electric Vehicle Usage*. SAE paper no. 2010-01-1040. Presented at the 2010 SAE World Congress, Detroit, MI.

Chaffin, D. B. 2001. *Digital Human Modeling for Vehicle and Workplace Design*. ISBN: 978-0-7680-0687-2. Warrendale, PA: SAE International.

Chaffin, D. B. 2007. Human Motion Simulation for Vehicle and Workplace Design: Research Articles. *Human Factors in Ergonomics & Manufacturing*, 17(5): , p 475–484.

Fai, T. C. and M. Rauterberg. 2007. Vehicle Seat Design: State of the Art and Recent Developments. *Proceedings of the World Engineering Congress*, 51–61, Penang, Malaysia.

Fillyaw, C., J. Friedman and S. M. Prabhu. 2008. *Testing Human Machine Interface (HMI) Rich Designs using Model-Based Design*. SAE paper no. 2008-01-1052. Paper presented at the 2008 SAE World Congress, Detroit, MI.

Human Solutions. 2010. *RAMSIS Model Applications*. www.human-solutions.com/automot ive/index_en.php (accessed: March 17, 2024)

Klauer, S., T. Dingus, V. Neale, J. Sudweeks and D. Ramsey. 2006. The Impact of Driver Inattention on Near/Crash Risk: An Analysis Using the 100-Car Naturalistic Driving Study Data. (DOT HS 810 594). Washington, DC: U.S. Department of Transportation, National Highway Traffic Safety Administration (NHTSA).

Parkes, A. M. and S. Franzen (Eds.). 1993. *Driving Future Vehicles*. ISBN: 978-0-7484-0042-3. UK: CRC Press.

Prefix Corporation. 2024. *Programmable Vehicle Model (PV)*. www.prefix.com/design-engi neering/ (Accessed March 17, 2024)

Reed, M. P., M. B. Parkinson and D. B. Chaffin. 2003. A new approach to modeling driver reach. Technical Paper 2003-01-0587. *SAE Transactions: Journal of Passenger Cars – Mechanical Systems*, 112: 709–718.

Reed, M. P., R. W. Roe and L. W. Schneider. 1999. Design and Development of the ASPECT Manikin. Technical Paper 990963. *SAE Transactions: Journal of Passenger Cars,* 108.

Regan, M. A., J. D. Lee and K. L.Young (Eds.). 2009. *Driver Distraction: Theory, Effects, and Mitigation*. Boca Raton, FL: CRC Press.

Sayer, J. R., J. M. Devonshire and C. A. Flanagan. 2007. Naturalistic Driving Performance During Secondary Tasks. *Proceedings of the Fourth International Driving Symposium on Human Factors in Driver Assessment, Training and Vehicle Design*.

26 Product Liability Litigations and Ergonomic Considerations

INTRODUCTION

Product manufacturers are sued in large numbers by users, misusers, and even abusers of their products. Injuries resulting from the use (or often misuse) are the basis for an increasing number of product liability lawsuits (Hammer, 1980; Goetsch, 2007). These suits cost the industries millions of dollars each year. Therefore, the objectives of this chapter are (a) to provide the reader background into basic concepts and issues related to product liability; (b) to provide an understanding into the role and importance of safety analyses in product and systems design; and (c) to help a product engineer communicate with a product liability lawyer.

The best way a manufacturer can prevent or defend such claims is by manufacturing a reasonably safe and reliable product (or system), and by providing warning labels and instructions for its proper use. The key to achieving a reasonably safe and reliable product and to reduce the product liability exposure is to "build-in" product safety during the early design stages of the product.

TERMS AND PRINCIPLES USED IN PRODUCT LITIGATIONS

The following terms used in product litigations will help the reader understand the issues (Hammer, 1980; Goetsch, 2007).

1. *Liability*: This can be defined as an obligation to rectify or to recompense for any injury or damage for which the liable person has been held responsible or for the failure of a person to meet a warranty. Here, the product user is the loser (or injured) and is assumed to be demanding compensation for the injury, losses, and/or suffering caused by the product.
2. *Plaintiff*: A person (or a party) who starts a legal case against another to obtain remedy for an injury caused by the product.
3. *Defendant*: A manufacturer (or the seller) who is faced with proving that the product is safe.
4. *Three major legal principles*: The following three principles are generally considered in establishing liability: (a) negligence; (b) strict liability; and (c) breach of warranty (Hammer, 1980).

DOI: 10.1201/9781003485605-13

a. The negligence principle involves failure of the defendant to exercise a reasonable amount of care in the design and manufacture of his product (or to carry out a legal duty) so that injury or property damage do not occur to a user or other person. Thus, here the focus is on the conduct of the defendant (manufacturer), that is, his duty and/or care. The plaintiff must prove that the defendant's conduct involved (i) an unreasonably great risk of causing damage; (ii) defendant failed to exercise ordinary or reasonable care; and/or (iii) not using available knowledge (e.g., new developments, design methods, safety practices, and safety devices) that would have decreased the level of risk.

b. The principle of strict liability is based on the concept that a manufacturer of a product is liable for injury due to a defect, without the necessity for a plaintiff to show negligence or fault. Here, the plaintiff must only prove that the product was defective, unreasonably dangerous, and the proximate cause of the harm. Thus, in strict liability, the focus is on the quality of the product – rather than the fault of the manufacturer – that is, regardless of whether or not the manufacturer acted reasonably. The manufacturer is said to be "strictly liable" because his liability does not depend on his own conduct or care. Therefore, defense is particularly difficult and frustrating to the manufacturer. This is the basic cause for what is called the product liability crisis.

c. The breach of warranty can involve the following two principles:

(i) implied warranty and (ii) expressed warranty and misrepresentation. The principle of an implied warranty involves an implication by a manufacturer or dealer that a product is suitable for a specific purpose or use, is in good condition, or is safe by placing it on sale. The implied warranty of safety is the principle that any product by being placed on sale is implied to be safe. The implied warranty of merchantability implies that the product sold is in as good a condition as other products of its type. The implied warranty for fitness implies that the product is suitable for the purpose for which it is sold.

The principle of expressed warranty involves a statement by a manufacturer or dealer, either in writing or orally, that his product will perform in a specific way, is suitable for a specific purpose or contains specific safeguards.

PRODUCT DEFECTS

The first step in product liability (in cases where no express warranty or misrepresentation is involved) is to prove that the product was defective. (Note: It is not sufficient to establish that the product was dangerous: e.g., even a knife can be dangerous). Thus, the product must have a design defect (i.e., a defect in its basic design) or a manufacturing defect (e.g., a flaw in the manufacturing process).

Examples of design defects are as follows: (1) a concealed danger created by the design, e.g., a sharp edge after collapse in an accident; (2) needed safety devices not included; and (3) involved materials of inadequate strength or failed to comply with accepted standards.

Examples of manufacturing defects are as follows: (1) poor quality material used for structural components; (2) failed to meet required material hardness (e.g., failure in the heat treating process); (3) a sharp edge or flash left on a grasp handle (e.g., operator forgot to grind the sharp edge, which can lead to an injury); and (4) improper assembly (e.g., misaligned parts, loose parts, missing parts, wrong electrical connections – transposed wires).

Warnings

The courts have recognized the concept of the product manufacturer's duty to warn the users about potential safety-related considerations and problems. The general position of the courts is to always use warnings. The cost of supplying warnings is low, as generally only the printing costs are involved (e.g., warning labels, warning messages included in display screens or user's manuals). In many cases, the manufacturers have been found to be liable for failure to warn, even when the (missing) warning would have been of dubious value. From the ergonomic perspective, too many warnings are ineffective as people will disregard frequent occurrences of warnings. But courts view the provision of warnings as desirable and thus it may have some effect on reducing the manufacturer's negligence (due to failure to warn). Thus, presence or absence of warnings is an important issue as compared with their effectiveness.

Engineering and management, thus, can be vulnerable in the following areas: (1) product design; (2) product manufacturing and materials selection; (3) packaging, installation, and application/use of the product (i.e., operation); and (4) warning labels. Failure to comply with the regulations (i.e., applicable standards) can mean that the manufacturer is negligent. Compliance with government standards (which are generally minimum standards and hence not very stringent) may provide some protection against negligence-related cases but it offers no protection in strict liability cases.

Safety Costs

Safety costs are all costs associated in incorporating safety in the product design and ensuring that the products operate safely during their operational life. These costs include the following:

1. Costs incurred in creating safe products:
   a. Costs associated in gathering data on safety regulations, past accidents, and litigations-related data.
   b. Costs associated with developing safety requirements, cascading requirements to various systems, subsystems, and components of the product (Note: Safety is a product attribute).
   c. Costs to design and implement safety features in the product (e.g., costs associated in conducting safety analyses, product safety design reviews,

meetings with experts, management, government agency experts, and product safety lawyers).

   d. Costs associated in following special safety precautions (e.g., checks, inspections) during manufacturing and assembling.

   e. Costs of verification and validation tests (i.e., safety testing costs).

2. Costs incurred during product uses/operations.

   a. Costs associated with gathering data on safety-related incidences (e.g., meeting with customers, users, dealers, repair shop personnel, government agency personnel, and lawyers, and investigating product failures and accidents).

   b. Conducting safety analyses and tests.

   c. Providing technical and legal support on product litigations, recalls, repairs, fines, customer relations campaigns, and so forth.

   d. Costs associated with fixing the defects (i.e., product recalls, repairs, retests).

3. Safety costs related to product discontinuation and disposal.

   a. Disposal or recycling of retired products.

   b. Disposal of plant equipment and hazardous/toxic substances (e.g., toxicity tests after disposal).

CASE STUDIES WITH ERGONOMIC CONSIDERATIONS

1. *Lamp Intensity Deviations*: The photometric outputs of automotive exterior lamps vary due to variations in light sources, reflectors and lens. Since each lamp has many test points at which the lamp output needs to be within specified minimum and maximum light intensity (candela) and chromaticity (color) coordinate requirements (see Chapter 9), the lamps fail to pass the FMVSS 108 requirements. In such situations, the lighting engineers consult ergonomics engineers to determine whether the photometric failures will be inconsequential or the lamps need to be recalled and brought under compliance. The ergonomics engineers may conduct additional visibility analyses and evaluation to establish any safety issues (e.g., correct recognition of a signal) that may result in accidents and product liability cases.

2. *Wiper Blade Failures*: Some vehicles have had wiper blades or wipers falling off during driving in rain. Drivers who have experienced such failures find it quite a traumatic experience especially in heavy rain. However, such wiper failures generally do not cause accidents and drivers have been able to safely stop their vehicles on the side of the road without reporting any accidents. Such incidences result in vehicle recalls fixing the wipers.

3. *Inside Door Opening Handles*: The locations inside door opening handles in a few vehicles have been placed high (but just below the beltline). There have been some cases where the unbelted front passenger falls out of the vehicle in sudden steering maneuvers (initiated by the driver) when the passenger

accidently grabs the door opening handle (instead of the inside door grab handle) resulting in opening of the door. The manufacturer can be sued for placing the door opening handle and shaping the handle such that in an emergency it can be mistaken for a door grab handle.

4. *Power Window Switches*: The power window switches in some older vehicles were equipped with momentary rocker switches (in the armrest of the driver's door) such that when the front end of the rocker is pushed the window will roll up and the window will roll down when the rear part of the switch is pushed down. In such vehicles, unattended children sitting in the driver's seat climbed up onto the door armrest, leaned out the driver's side window, and stepped on the front part of the window switch. The activation of the switch resulted in raising the window glass up and strangulating the child. Such puh-push momentary rocker switches (push down the front end to raise the window and push down the rear end to lower the window glass) are now disallowed and replaced with pull up the front part of the switch to raise the window up and push down the rear part of the switch to lower the window. This change in switch design has reduced child strangulation injuries and the resulting liability cases.

5. *Ignition Key Switches*: A major vehicle manufacturer had a faulty design of the ignition key switch which turned off the key switch when the vehicle went over a pothole (due sudden shift of weight of other keys hanging on the ignition key with an oblong hole in the key head and the key switch with lower key turning torque). The turning of the ignition key switch resulted in suddenly and simultaneously turning off (a) the engine; (b) the power steering mechanism; and (c) the driver's air bag inflating mechanism. Such situations created serious and fatal accidents resulting in a recall to replace the ignition switch with a new key and key switch design.

6. *Electronic Gear Shifter*: A major manufacturer designed and installed a new monostable console mounted electronic gear shifter. The shifter was designed such that it returned back to its neutral shift position (after the driver made the intended shift movement) without providing any tactile feedback to the driver. The shifter mechanism did provide a lighted setting label (to indicate selected gear position) on the shift knob and also a lighted gear shift indicator in the instrument cluster. However, drivers who do not view the lighted shift position indicator did not confirm their shift position and ended up leaving the vehicle in reverse gear instead of in the park position. There were a number of cases where the vehicle rolled under power in reverse and ran over the driver trying to stop the vehicle.

7. *Burn Injuries from Hot Vehicle Surfaces*: Some vehicle models were produced without sufficient insulation or heat shields to reduce temperatures of vehicle surfaces in the interior of the occupant compartment and/or cargo box of pickup trucks. The hot catalytic converter in the exhaust system attached under the vehicle floor increased temperatures of some body parts over 140^0 F causing instantaneous skin burns. Unknowingly touching such hot body parts caused severe burn injuries which resulted in liability cases.

CONCLUDING REMARKS

Product safety is an important discipline, and the product designers must make sure that their products are safe. Safety requirements must be incorporated early in the product design and product attribute engineering process. Safety reviews must be conducted during both the design and manufacturing phases to ensure that the products do not have any design and manufacturing defects that will increase risks to the users as well as the manufacturer. It should be realized that during the product liability cases, the manufacturer is considered to be an expert and knowledgeable about (a) the product safety requirements; (b) available safety devices; (c) safety technologies; and (d) safety-related design and manufacturing considerations. Furthermore, the manufacturer is also expected to provide warnings to the users about any potential hazard associated with the use of the product. The design engineers should maintain proper records on safety analyses and safety-related decisions based on potential benefits of the product to the customers versus costs incurred to make the product reasonably safe. The records will be useful in defending their decisions if challenged during any future product reviews and/or liability cases.

REFERENCES

Brown, D. B. 1976. *Systems Analysis and Design for Safety-Safety Systems Engineering.* Englewood Cliffs, NJ: Prentice-Hall, Inc.

Colling, D. A. 1990. *Industrial Safety Management and Technology.* Englewood Cliffs, NJ: Prentice Hall.

Goetsch, D. L. 2007. *Occupational Safety and Health for Technologist, Engineers, and Managers.* Sixth Edition. ISBN: 0132397609. Englewood Cliffs, NJ: Prentice Hall.

Hammer, W. 1980. *Product Safety Management and Engineering.* Englewood Cliffs, NJ: Prentice-Hall, Inc.

Hammer, W. 1989. *Occupational Safety Management and Engineering.* Fourth Edition. Englewood Cliffs, NJ: Prentice Hall.

Heinrich, H. W., D. Petersen and N. Roos. 1980. *Industrial Accident Prevention.* Fifth Edition. New York: McGraw-Hill, Inc.

National Safety Council (NSC). 2012. *Injury Facts.* 2012 Edition. Spring Lake Drive, Itasca, IL: National Safety Council 1121. http://shop.nsc.org/Reference-Injury-Facts-2012-Book-P124.aspx (accessed October 25, 2012).

Index

Note: Page numbers provide volume number, hyphen and page number (or range of page numbers) in the volume.

Printed in the United States
by Baker & Taylor Publisher Services